Conservation Science
Heritage Materials

Conservation Science
Heritage Materials

Edited by

Eric May
*School of Biological Sciences, University of Portsmouth,
Portsmouth, UK*

Mark Jones
The Mary Rose Trust, HM Naval Base, Portsmouth, UK

RSCPublishing

ISBN-10: 0-85404-659-3
ISBN-13: 978-0-85404-659-1

A catalogue record for this book is available from the British Library

Published by The Royal Society of Chemistry,
Thomas Graham House, Science Park, Milton Road,
Cambridge CB4 0WF, UK

Registered Charity Number 207890

For further information see our web site at www.rsc.org

Typeset by Macmillan India Ltd, Bangalore, India
Printed by Henry Ling Ltd, Dorchester, Dorset, UK

Preface

Conservation science is a broad church and a rapidly developing discipline. The aim of this book is to provide, in one volume, an account of the composition and conservation of historic materials. It brings together recent information on decay and degradation that are unusually scattered in journals and not readily available to many practising conservators. By doing this, it is hoped that the potential of materials conservation, together with the processes that are necessary to maximise it, can be made more understandable to all those interested in historic objects. We believe that a clear understanding of the deterioration mechanisms of historic and artistic works will help conservators make informed decisions about their care and treatment.

At the time of writing, there is concern throughout Europe about the fragmented nature of research funding in conservation science, despite the fact that heritage generates considerable income through tourism for national economies. There is parliamentary activity at national and EU levels as we move to the 7th European Research Framework. This book demonstrates that a very wide range of accumulated deterioration problems affect materials that are part of our movable and immovable heritage. The threat of environmental climate change now compounds those problems and seriously extends the scale and nature of the challenge that conservators and conservation scientists face. The wealth of knowledge offered by specialists in this book illustrates the commitment to good science and reinforces the need for on-going research to meet the new challenges.

The book is intended for graduates and professionals who are involved in the understanding and care of historic materials, whether conservators, finds specialists, museum curators or administrators. It is also of interest to undergraduates studying archaeological conservation at university, or scientists taking subsidiary courses in conservation. It should also prove useful to those involved in the training of students in material conservation. Inevitably, when describing deterioration (decay), material composition and treatment methods, science is essential for the explanation. However, the general reader with basic science knowledge should have no difficulty in following the text. It should provide a good introduction to the subject and we hope that it will be a valuable reference work and precious reading companion for both students and professionals.

The text is divided into chapters covering analytical aspects, organic materials, inorganic materials, wall paintings and *in situ* preservation. Contributions are from a wide range of practising international experts from leading research institutes who outline the problems and provide many interesting and exciting case studies. The text offers a basic introduction to the science of heritage materials, their deterioration and restoration processes. The complexities and problems faced by conservators and scientists are addressed together with the processes of decay and degradation that affect ancient materials. It describes in detail the processes of decay and degradation for each major category of material (paper, wood, stone, glass, metals, leather, plastics *etc.*). The chapters on plastics and *in situ* conservation provide insights into the rapidly developing new areas of conservation science.

A word here about topic coverage and our policy on referencing. These are generally full with topics where there exists no adequate books or review articles – metals and wood are cases in point – and, as a consequence, there is no necessary correlation between length of a chapter and the practical importance of the subject. In relation to the references, we have adopted a non-referenced style and provided reference sources and further reading at the end.

This book is clearly the product of many specialists and the editors are conscious that without their input it would not exist. It has had a very long gestation period. There is never a good time to edit a book; it is a labour of love that surpasses all understanding and common sense, measured against the increasing pressures of academic life. The contributions have come in over a long period and in that sense we are grateful for the understanding of not only the authors, but the RSC, particularly Janet Freshwater and Annie Jacob, who have cajoled, but mostly supported, us throughout a prolonged delivery. We are also grateful for assistance and help from our organisations, namely, the Mary Rose Trust (MJ) (notably Andy Elkerton with the index) and the University of Portsmouth (EM). This could not have happened without the support of our families, especially Helen and Jeannette. We hope you enjoy the final product.

Eric May
Mark Jones
May 2006

Contents

CHAPTER 1

Introduction

MARGARET RULE

Maritime Archaeologist; Archaeological Director, *Mary Rose* Project 1967–1994; Archaeological Consultant 1995–2003

1 CULTURAL HERITAGE: THE ARCHAEOLOGIST, THE CONSERVATOR AND THE PUBLIC – AN ESSENTIAL COALITION

The recent growth of public interest in archaeology and their enthusiasm for the past is reflected by the astonishing success of 'heritage' programmes on television. A stranger to British television is often surprised at the number of cookery, gardening and archaeology programmes offered for entertainment every day and their success may reflect a growing wish to escape from the stresses of modern living and a desire to understand our past. All three subjects can be related to our heritage and the wonderful legacy left by our ancestors.

The better archaeological programmes often demonstrate the application of technology to solve the problems of dating, identifying and conserving cultural material, and the range of periods, sites and cultures investigated is wide and diverse. Usually, these techniques are not new as many were originally developed for commercial or military use, but their application to archaeology is new, and the public, through good television presentation, is able to understand both the technique and the problem.

One evening, we joined the architectural historians as they solve the problems of how a building developed as its use changed over the years, and the consultants discussed how best to interpret and preserve it for the future. The next evening, we joined a group of students of garden history who explained the importance of soil analysis and remote prospection using ground-penetrating radar or a magnetometer to enable the researcher to present an accurate recreation of a lost 18th-century garden. In the realm of the Field Archaeologist, we are offered programmes several times a week showing desperate attempts to solve all the problems of a complex excavation in 72 hours.

The use of modern computer techniques and the abilities of the forensic scientist skilfully displayed within these programmes have familiarised the general public with technology in a way that has never happened before. The result is that they cannot be fobbed off with conjecture about archaeological material – they expect facts. They also expect that the objects and any evidence found by modern archaeologists will be preserved to be reassessed and displayed for education and pleasure. In parallel with the growth in public interest there has been a growing investment of public money and with this comes accountability.

1.1   The Archaeologist

The task that is faced by the archaeologists, whether their work is publicly funded or not, is enormous and the skills required are diverse. The recognition that any excavation of buried material is inevitably destructive means that the intrusion into a site has to be justified and well planned. Pre-disturbance appraisal including soil analysis, geo-prospection, topographical survey and aerial photography will provide the outline for a strategic plan and a wise archaeologist would include the scientists in planning the programme of work at this early stage.

Archaeology has moved away from the collection of individual 'things' to the collection of information about the people who made and used the objects. This change in attitude requires a forensic approach and commitment far removed from treasure-hunting and early barrow digging. Even an important and highly glamourous site like the 7th century tomb discovered at Prittlewell, during the rescue excavations in advance of road development, was part of a well-designed programme founded on examination of the material discovered in the 1920s. The importance of the site, with its mixture of Christian and pagan ritual material, cannot be over emphasised and although the public may be dazzled by the gold crosses, the presence of wooden objects and small bone dice suggest that soil analysis may yield more information. The lost burial and vegetable materials buried with the body may throw new light on the environment in Essex in the early 7th century; only time and rigorous analysis of samples will tell.

1.2   The Conservator

The skills of the conservator and the high standards of diagnostic analysis and documentation required are an essential part of any programme to investigate and preserve evidence of our past. Whenever possible the monument or building should be preserved *in situ* in the context and the landscape where it was created. Interest in stately homes and public buildings exceeds interest shown in a sequence of objects displayed as art objects or collector's items in a museum gallery. The kitchens and the bathing arrangements in a house like Longleat, Wiltshire, UK are at least as interesting as the paintings and ceramics. However, the collections in a stately home need constant vigilance to ensure that climate changes, insects and dust do not destroy the legacy. While the house itself, like any building, needs constant repair to ensure that it will survive.

Any homeowner visiting a house like Longleat will recognise the problem but the scale of the maintenance and repair programme on any building whether it is a Cathedral, a Parish Church or a tollbooth is large and the needs are constant and unavoidable. This cannot be entirely funded from public funds and good interpretation and user-friendly display are essential if the site is to earn enough income to support the project. Both commercial sponsors and the public need to know that the site is well managed and that the future of the building and its contents is secure.

2 FROM DISCOVERY TO DISPLAY

So the archaeologist has to have a plan that will ensure good management from discovery to dissemination of information by popular and scientific publications and by public display. He or she also needs good scientific advisors. Certain constraints remain as tried and tested dogma:

1. Sites should not be excavated unless they are threatened with destruction or they fit into an agreed programme of research.
2. All excavation should be preceded by non-destructive survey.
3. A full programme of investigation, excavation, conservation, scientific analysis and publication must be agreed before work starts.
4. Funding should be in place or guaranteed from the start.
5. All stabilisation and preservation treatments should be reversible.

While all these constraints are desirable it is rare for them all to be possible before work begins and any programme must be monitored and reassessed as work progresses. The importance of involving a multi-disciplined team in the process of establishing and monitoring any programme cannot be overemphasised. Usually the archaeologist has to divide a programme into phases and progress from one phase to another depends on the results of one phase and the availability of funds.

Often a site is partially destroyed before the archaeologist is involved. This occasionally happens on land sites if there is poor liaison between the archaeologists, the planners and the developers. This happens more often underwater where a site may be unexpectedly exposed by unusual storms or a change in currents caused by the construction of sea defences or offshore dredging.

2.1   Special Problems of Underwater Recovery

The usual scenario is that the exposed wreck site is discovered by divers, either accidentally or as a result of research. The team then try to identify and record the wreck and some objects are recovered to aid the process of identification and brought ashore. At some stage the finds and the site are reported; in the UK this would be to the Receiver of Wreck and eventually to English Heritage. If the site is thought to be of importance a team of professional diving archaeologists will visit the site, eventually the site may be designated as being of importance under the Protection of Wrecks Act 1973.

If the original team of divers is lucky they may have the unpaid assistance of professional archaeologists and conservators who will advise them from an early stage. They may also decide to apply for their 'wreck' to be designated with the approval of their team as licensees. A programme of work has to be submitted before the application for a licence will be approved and among other requirements there is the requirement for a nominated archaeologist and a conservator. A difficulty arises if the team (or the Ministry) requires the archaeologist to work full-time alongside the amateur team whenever the amateur team is on site. That would mean that the archaeologist would have to be paid and, as they would then be a diver at work, they would need professional diving qualifications, insurance and the support of professional divers as backup. The responsibilities and the costs rise beyond the budget of a team of vocational diving archaeologists. In the face of this it is inevitable that some sites will not be reported.

2.2   The Raising of the *Mary Rose*: A Case Study

The *Mary Rose* was discovered (see Figures 1 and 2), surveyed, excavated and recovered using a pre-designed floating programme of work that could be adjusted and cut if circumstances changed. It was recognised that any disturbance of the seabed would threaten the physical integrity of the wreck and plans were made accordingly. From the beginning of the search phase in 1967, there was an archaeologist in the team and advice on conservation was always available on request from Portsmouth City Museum and scientists at Portsmouth Polytechnic (now the University of Portsmouth).

Objects, including large guns, were recovered from the sea prior to the formation of the *Mary Rose* Trust in 1979, and they were taken to the conservation laboratories of the City Museum the same day. Close planning and liaison with the people providing equipment to lift the guns from the wreck site and take them ashore, the road transport and the museum staff were essential. Once the operation became full-time in 1979, the *Mary Rose* Trust, a charity and a company limited by guarantee, was formed to excavate and, if feasible and

Figure 1 *Frame P 24. The first frame discovered in 1971*

Figure 2 *P2 in 1974. Biological degradation has severely eroded timber exposed above the mud, but the migration of iron from the survey nail has inhibited deterioration in the iron rich zone around the nail*

desirable, recover the wreck of the *Mary Rose* and all its contents. There were a series of cut-off points, and these were essential because there were no funds available from the local government. The whole operation depended on commercial support for equipment and personnel and the generosity of the public.

A full-time team of divers, professional diving archaeologists and conservators were recruited to do the fieldwork and a small team including a press officer, scientists, illustrators and conservators worked on the finds once they were ashore.

A programme of sampling the sediments that had begun in 1977 was enhanced and an environmental scientist was employed by the Trust in 1979 to analyse the material and select samples for appraisal by consultants. The strategy of the sampling programme was two pronged:

> First – To identify environmental material derived from containers including barrels and boxes or as contaminants on clothing, cordage and dunnage.
> Second – To identify a series of microenvironments within and around the ship.

This was necessary in order to understand the complex processes of sedimentation, preservation and collapse, which occurred immediately after the ship sank and the series of events that occurred after the wreck was buried and localised intrusions occurred (see Figure 3).

Figure 3 *Silts between decks on the* Mary Rose. *The fine sediments that filled most of the lower hull had never been disturbed, and in this anaerobic environment organic materials such as silk, leather and wood were well-preserved*

For the *Mary Rose*, the cut-off points were designed to present the least possible threat to the ship if the project had to be aborted. In the early years, the site was backfilled at the end of each season. As work progressed, the *Mary Rose* site became too large (roughly $40 \times 25 \times 3$ m) and the work involved in backfilling and re-excavating the site would have been counter-productive. Therefore the exposed timbers of the hull were draped in Terram, an inert geo-textile, held closely in position by carefully chosen exotic gravel that could be easily recognised and removed if work continued in the following year. This cover protected the hull timbers from mechanical attrition by current-borne detritus and inhibited colonisation of the timbers by larger marine animals and plants.

Only after the decision to lift the hull was made in January 1982, was the final excavation begun to undercut the hull and leave the empty hull for recovery the following autumn. Work to define the engineering strategy to recover the ship had begun two years before, and in the spring of 1982, work began to prepare the ship for recovery and conservation.

A conservation laboratory equipped to deal with finds had functioned at the Trust headquarters at Portsmouth since 1979, and the external Advisory Committees met regularly to review progress and monitor the work as it progressed. In 1982 a suitable dock in the Royal Naval Base was equipped to receive the hull while it was cleaned, restored and conserved. The salvage plans and a detailed research programme to define passive and active conservation techniques went on concurrently with the excavation offshore between 1980 and 1982.

2.3 Preservation of the Evidence: A Multi-discipline Task

Organic material recovered from the sea, where it has been buried in a relatively stable anaerobic environment, is often well-preserved with fine detail. The obligation to preserve and record the evidence during excavation is paramount. Even on land sites, archaeologists have to contend with natural and man-made threats during excavation, and temporary shelters from rain, wind and frost are often necessary. Trying to excavate and record the remains of a timber building during a typical English summer is sometimes heart-breaking. The evidence is usually preserved as a series of changes in colour and texture and if the natural subsoil is clay or brick–earth, then it is often difficult to record and there is little to distinguish between the remains of a clay and daub structure and the natural subsoil. Summer weather is unpredictable and one day the archaeologist will be using a garden syringe to spray the site and reveal the features for photography while next day he will be erecting scaffold supports and tarpaulins to protect the site from a deluge of rain. In England it is often bad enough but in tropical regions a site that has been a desert for weeks may be destroyed and washed away completely by heavy rain.

However, underwater the problems are magnified. Tidal changes, storms and contamination by anchorage debris have to be contended with while an excavation is in progress. Decisions to leave an object *in situ* until it has been surveyed and photographed properly have to be resisted. The archaeologist, whether professional or amateur, should be capable of fixing the location of his find by reference to pre-determined datum points at the time of discovery and he should always carry a camera and a scale. This is not to advocate a policy of snatch and run, but an object or a sample swept away by a spring tide is of no use to anyone. Working in the turbulent shallow waters of the reefs of the Cape Verde Islands, I learnt that everyone, professional or amateur, had to be trained to record as they worked. There was often no going back; at least until the weather changed.

A supply of small sandbags stored close to the excavation are useful to support or protect an area while seeking additional help or advice but leaving an object exposed to tidal changes is a matter of good judgment and experience. Because there is no rule of thumb about when to leave and when to lift, it is essential that an experienced member of the conservation team should be a diver. The senior conservator should be aware of the situation underwater and be prepared to evaluate the natural threats to the underwater deposits, ensuring that each diver recognised the problems and was properly briefed and equipped before each dive. Preservation and documentation begin at the moment of discovery.

Whenever possible, objects should be lifted in a rigid container (not in a bag) and packed with a representative sample of the surrounding matrix to limit movement and damage during recovery. A supply of numbered plastic boxes with attached lids should, like the sandbags, be stored close to the excavation. If surface conditions are severe, the containers, suitably numbered and recorded, can be stored on the bottom until calmer conditions occur.

Backfilling, to protect a site between working periods requires judgement, experience and a thorough knowledge of seasonal changes in the area around the wreck. Once again there is no rule that applies to all sites. Knowledge of the site and an evaluation of any potential hazards, which might threaten preservation, are essential. Discussion with local fishermen whose lives depend on their knowledge of the local area can be very rewarding, particularly in parts of the world where meteorological data is difficult to obtain. Tide Tables are a useful guide, but it is more useful to calibrate any variations for a specific site and produce modified and reliable data. In many parts of the world, tables do not exist or they were drawn up in the dreamtime. It is easy to construct a tide gauge on a nearby beach and monitor the changes through a lunar month. The golden rule is that preservation, *i.e.* protection, begins at the moment of discovery and not when the object is sent ashore to the laboratory.

In Guernsey, Channel Islands, the remains of a 3rd century merchant ship lying in shallow water (4.8 m below chart datum) were covered with 16 tons of sandbags to protect the timbers. They were swept away in one storm leaving the site dangerously exposed (Rule and Monaghan, 1993). Thankfully, volunteer divers braved the winter weather to replace them, building the mound carefully like a Flemish bond brick wall. Their care saved the timbers and they are now awaiting display in St Peter Port.

Loose timbers that were swept away from this wreck soon after exposure in 1984 were attacked by gribble and in 12 months all surface detail was lost. After 24 months exposed loose timbers had lost 50% of their bulk and were virtually unrecognisable (see Figure 4).

In the early days of the *Mary Rose* volunteers did the work. They were dedicated divers who had searched for the wreck and who limited excavation to minimum disturbance. The ship and her contents had been preserved in mud, sand and silts that provided an anaerobic environment and inhibited micro- and macro-biological degradation. One of the major problems was caused by infestation with *Limnoria* sp. To prevent or limit this infestation we backfilled the area with the mud removed during the excavation. This was only moderately successful. The upper levels of mud in this area of the Solent are not clean and it was rather like throwing a dirty cocktail of mud and effluent over the cleaned and newly exposed timbers. After many discussions with scientists at the University of Portsmouth and Imperial College, London we chose to blanket the area with hessian covered with sandbags. 1000-gauge Polythene, the first choice, proved to be too difficult to handle underwater.

Figure 4 *Detached timbers eroded by teredo in the upper levels of sediment below the stern castle* (Mary Rose)

Later, in 1980 when Terram, an inert blanket of geo-textile, became available we used it to clad all the exposed timbers. Any pockets in the blanket provided homes for squat lobsters and damage to the blanket allowed irrigation and inevitably more *Limnoria* infestation. The problem was to secure the textile closely to a three-dimensional structure. The ship was open as a wreck. Ribs, planking, deck beams and knees were exposed *in situ*. There were no flat surfaces. It took dedication and hard work to wrap the exposed surfaces with the Terram. To secure the blanket over the winter 1980/1981 we filled the excavation with a washed gravel aggregate. This was supposed to be 50 mm in diameter, but many were 100 mm. It was time-consuming to remove, but unmistakable, and it reflected the light well, an important point when underwater visibility was often reduced to 1 m.

In other parts of the world, a synthetic grass mat has been used to collect current-borne sediments enabling a protective mat of sand to accumulate. In Boa Vista, Cape Verde Islands the natural seabed is composed of water-worn pebbles of volcanic rock. These are easy to handle underwater and a layer of 100 mm pebbles overlain with some well-chosen boulders protects the site from marine and human interference. The insertion of a short tail of orange cord serves as buried marker. It is discrete, moving gently in the current, and to the casual intruder is much like other marine growth and debris.

What one cannot do is walk away from an underwater site and rely on your favourite deity to preserve it. By your intrusion into the environment of the wreck you will have jeopardised its chance of survival. The Guernsey wreck was nearly destroyed by being exposed by the propeller thrust of legitimate shipping using the harbour. In two or three years it would have been destroyed. Thankfully the amateur diving archaeologist who found the site received the support of his local club who were prepared to do anything asked of them. He also had the acumen to contact professional archaeologists, scientists and conservators who managed the project. This cooperation and respect within a multidiscipline team is essential.

The need for on-site conservators and recorders is obvious. Some details like paint marks on wood or stone are often fugitive and light labile, as were the markings on the parts of a seaman's chest from the *Mary Rose* indicating how they fitted together. Obviously it was a flat-pack kit purchased for later assembly. Within 20 minutes the marks had been photographed, but they were beginning to fade and there is no trace of them today.

3   CONSERVATION WITH A PURPOSE

The importance of conserving cultural material is obvious. The public expect it, and the specialist also expects to be able to reassess the evidence. The material should also, whenever possible, be available for study and exhibition away from its 'home' in a museum or library or storeroom. Environmental conditions for storage, transport and display should be defined, controlled and monitored.

Textiles, documents and paintings suffer irreversible damage if they are displayed in galleries with high light levels, and monitoring the ultra-violet levels is required. Metals recovered from the sea are often unstable especially if the environment in a gallery is too moist or fluctuates wildly as the external weather changes.

Gallery conditions required to mount a travelling exhibition are rigorous, and they need to be defined and agreed before any agreement is made between the owner and the exhibitor. Light levels, relative humidity and violent changes in room temperature are threats to delicate materials and all material, whether it is recovered from the sea or from a land site, is vulnerable if it is carelessly displayed or made available for handling by visitors or school parties and the media. The use of white cotton gloves is now *de rigueur*, but at one time the use of white gloves on a television programme brought howls from the producer as the might 'cause glare'. It is now expected.

4 MUSEUMS: THE PAST INTO THE FUTURE

If we recognise that cultural material is a non-renewable resource that cannot be replaced, we must preserve it. One way of avoiding the dangers of damage by ill-advised or careless exhibitors is to furnish a loan exhibition with a careful selection of stable material such as stone or ceramics illustrated with good photographs, video or film and supported by first class replicas. Replicas of metal objects, bone and glass can be made that faithfully recreate the form, colour and density of the original. They are so good that they have to be marked as replicas so as not to deceive the public.

It has been argued that the beauty of the Parthenon sculptures, now scattered across 10 museums in eight countries, could be brought together in Athens or the British Museum if every piece was scanned using laser technology to create a 3D virtual reconstruction. This would allow scholars to study the monument as it was, before it was partially destroyed in the early 19th century. The evidence would remain unaffected by climate change and international politics and, if required, full-scale replicas could be made using the accurate survey produced by the laser scans.

Expensive? Yes. Worthwhile? Ask the public who save up for years to go to Disneyland. Any popular use of good replicas for the purposes of education or entertainment helps to maintain the physical integrity of an irreplaceable object and the thrill a child gets handling and playing a unique 16th-century musical instrument is not diminished by the fact that it is a skilful replica.

These replicas, with accompanying contemporary music and details of the scientific methods used to record the original object and manufacture the replica, can travel the world. Meanwhile the original is safely displayed in a carefully monitored, controlled environment.

The relationship between the public, the conservation scientist and the archaeologist is symbiotic. Each group impacts on the work of the other. Mutual respect is the catalyst that will ensure that our cultural heritage has the future it deserves.

REFERENCES AND FURTHER READING

J.M. Cronyn, *The Elements of Archaeological Conservation*, Routledge, London, 1990.

M. Rule, *The* Mary Rose – *the Excavation and Raising of Henry VIII's Flagship*, Conway Maritime Press Ltd, London, 1982.

M. Rule and J.A. Monaghan, *Gallo-Roman Trading Vessel from Guernsey*, Guernsey Museum Monograph No.5, Guernsey, 1993, pp 6–7; pp 125–131.

J. Thomson, *Manual of Curatorship*. Butterworth-Heinmann, Sevenoaks, 1986.

CHAPTER 2

Methods in Conservation

A. ELENA CHAROLA[1] AND ROBERT J. KOESTLER[2]

[1] Scientific Advisor, Program Coordinator, World Monuments Fund-Portugal, Mosteiro dos Jerónimos, Praca do Império, 1400–206 Lisbon, Portugal
[2] Director, Museum Conservation Institute, The Smithsonian Institution, Museum Support Center, 4210 Silver Hill Road, Suitland, MD 20746 2863, USA

1 INTRODUCTION

Conservation is an interdisciplinary field that needs to take into account such disciplines as the humanities, the arts, the sciences and technology, and crafts. Although the actual conservation intervention is, or should be, carried out by trained conservators, the degree of that intervention cannot be decided by the conservator alone. The reason for this is that conservation is not merely a technical operation on a cultural object. Rather, it is a cultural operation carried out by technical means.

Thus, when a conservation intervention is being planned on an object, it is important that as much information as possible about that object be available to the conservator. For example, this includes the anamnesis of the object, *i.e.* its history, starting with its provenance, its historical context, including the available technology and materials available at the time the particular object was created, its subsequent use, and the conditions under which it was kept there.

It is clear, therefore, that the contribution of historians and art historians is critical for conservation. Furthermore, the study of the object, its constituent materials and any deterioration process that may be affecting it, requires the contribution of the sciences to effectively identify any changes and decay it may have suffered. Based upon the scientific information acquired, an appropriate conservation method can then be devised.

2 PRELIMINARY EXAMINATION

Any conservation intervention needs to follow a logical procedure. This starts with the visual assessment and the compilation of all the relevant historical data available, including its recent history and information on any previous conservation intervention. Then a diagnosis as to the state of conservation of the object is required. Is the object sound? Does it suffer from any deterioration? If so, what are the causes? These questions, as well as the identification of the constituent materials require the support from various analytical techniques that are detailed below.

The results from these analyses serve various purposes. In the first place, they have to correlate with the historic information available. For example, was the pigment identified on a ceramic or on a painting available at the time and place the object was supposed to have been manufactured? Was the technology used in its manufacture consistent with the available information regarding the culture that supposedly manufactured the object? Second, is the object in question in a sound condition or does it show some deterioration? If so, is that deterioration active or is it the remnant of a previous decay process that has now ceased?

These are fundamental questions that need to be addressed so as to determine if a conservation treatment is required or not; and should a treatment be necessary, the most appropriate method and material need to be determined. Finally, and as a result of this preliminary examination and study, a condition report needs to be prepared including any recommendations regarding subsequent treatment(s) or the condition(s) under which the object should be stored or displayed.

3 ANALYTICAL METHODS

The examination of an object starts first of all with a visual examination aided by the use of loupes and microscopes. Then, if necessary, other analytical techniques may be employed to study particular aspects of the object. There is no single analytical method that will provide all the answers needed. Depending on the object and the problem to be studied, different techniques need to be used. In many cases, two or more techniques have to be used to confirm the data obtained.

It is also important to remember that analytical techniques have varying degrees of sensitivity when it comes to detecting the presence of an element or compound. Thus, not finding a particular element may not necessarily mean that this element is not present in the sample, rather it may be that the technique used does not have the required sensitivity to detect it at a low concentration. The sensitivity of a method depends both on the method and element or compound to be detected. In many cases, it will also depend on the "matrix", *i.e.* the other compounds with which it is associated within the object.

The most useful techniques for conservation purposes are those that will identify the presence of a specific element, compound, or class of compounds, thus providing a qualitative analysis of the sample. In many cases, this analysis could give an idea of the relative concentrations of the elements or compounds found. However, as mentioned above, this is only an approximation and unless certain standards are used, the result can only be considered as semi-quantitative.

With the widespread computerisation of analytical instruments, results are often expressed as percentages, and, since a computer processes them, they are in most cases accepted as valid results by operators who do not have a thorough training in analytical techniques. It is therefore important to highlight that the interpretation of analytical results requires training and experience to be performed reliably. In particular, when dealing with archaeological and/or museum objects, which in general are unique, extensive experience is extremely important as the results obtained have to be interpreted in view of the historical information available about the object.

Most analytical techniques are based on the following principles: interaction of radiation with matter (such as Radiography, X-ray Diffraction (XRD), X-ray Fluorescence (XRF) and Fourier transform infrared spectrometers (FTIR)), or, interaction of elemental particles with matter (such as scanning electron microscope (SEM), energy dispersive X-ray spectroscopy (EDS), Thermoluminescent dating (TL) and Radiocarbon dating). More detailed information on some of the techniques described below and other special methods (such as UV and IR photography, gas chromatography and mass spectrometry) can be found elsewhere (Ferretti, 1993; Parrini, 1986; Sinclair *et al.*, 1997; Ainsworth, 1982).

3.1   Interaction of Radiation with Matter

Electromagnetic radiation has a broad spectrum, ranging from X-rays at high energies (short wavelength) to low-energy infrared light (long wavelength). The high-energy X-rays are capable of penetrating through solid bodies (hence, the need to be extremely cautious when using this radiation and the importance of having a good shielding system to protect operators). Since different materials have different densities, they will allow more or less radiation to go through. This is the principle used in radiography.

The high-energy X-rays when passing through matter are able to excite the electrons of the inner shells of an atom. When the excited electrons fall back to their original position they release the energy absorbed. This can be measured and is the principle used in X-ray fluorescence.

Radiation, when passing through an object that has "openings", that is, in the same order of magnitude as its wavelength(s), is subject to the physical phenomenon of diffraction. This is the result of interference, *i.e.* that some wavelengths will be enhanced and other cancelled. When dealing with visible light, this phenomenon is the one that gives rise to the different colours visible, for example, when a thin oil film forms above water, the colour of an opal or some butterfly wings. Since X-rays have much smaller wavelengths, the spacing required to produce diffraction are found in the crystal lattice of materials. This principle is used for the characterization of materials in X-ray diffraction.

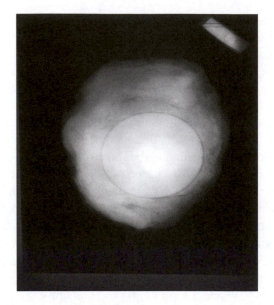

Figure 1 *X-ray radiograph of sixteenth century iron shot*

When dealing with low-energy infrared radiation, the interaction with matter is limited to the absorption of light by the outer shell electrons, *i.e.* those used in forming compounds. Hence, particular bonds will absorb particular wavelengths. This is the principle used for infrared spectroscopy. There are equivalent techniques for ultraviolet radiation and visible radiation, but they are mostly used to provide information about concentration of a given compound, rather than for identification purposes such as XRF or IR techniques.

Radiography.   Radiography of the object is based on the capacity of X-rays to pass through most solid objects and to blacken a photographic film. The amount of X-rays of a given energy passing through an object will be dependent on the thickness and density of the material of the object. In general, the X-ray source is placed above the object and this, in turn over a photographic film. Usually, an X-ray tube serves as the radiation source whose energy can be controlled by the voltage applied to the tube, to produce the radiation. This will vary from 5 to 30 kV for drawings and paintings on wood, to between 250 and 1000 kV for large bronze and stone statues. The other variable that can be controlled is exposure time. Optimizing the conditions to obtain a clear radiograph of an object is essentially an empirical operation (see Figure 1). Radiography will allow one to understand much of an objects manufacturing technology as well as allowing one to see any previous structural interventions.

X-Ray Fluorescence.   XRF will only provide information about the elemental composition of a sample (see Figure 2). This will be mostly limited to the

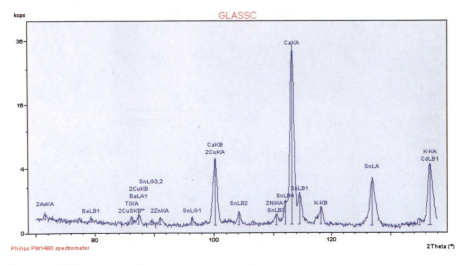

Figure 2 *XRF spectrum of heavy metal coating of glass*

surface of the object. The analysis is carried out in an air-path, or sometimes in a helium-flushed environment. One of the advantages of this technique is that it can be performed on large objects without sampling. However, the weaknesses are that, in an air-path, elemental detection is limited to the elements silicon and those with higher atomic numbers. Given that several layers can be present on the object, the depth of penetration of the X-ray beam may cause elemental information to be collected from different layers. In particular, this hampers quantitative analysis, the results being at best semi-quantitative if standard materials of similar composition are available for comparison. The detection limit for this method can be in the order of tens of ppm (μg/g).

X-Ray Diffraction.   When monochromatic X-rays impinge on a crystalline material in which the crystal lattice dimensions are in the order of the wavelength of the X-rays, diffraction of the beam occurs. This is the result of the physical phenomenon of constructive (or destructive) interference. As a result, a diffraction pattern emerges where some beams are reinforced and other cancelled (see Figure 3).

   XRD is generally carried out on finely powdered materials that are spread uniformly on an amorphous silica petrographic slide that is inserted into an isolated unit and then irradiated by monochromatic X-rays (in general the CuKα radiation). The detector is rotated with regards to the impinging X-ray beam, at a given speed. The diffractogram obtained is a plot of the intensity of the diffracted X-rays as a function of the angle (2θ) formed by the detector and the impinging X-ray beam. The finely powdered material, randomly oriented on the slide ensures statistical probability of obtaining a correct diffraction pattern of the material. The pattern obtained is compared with those of standard

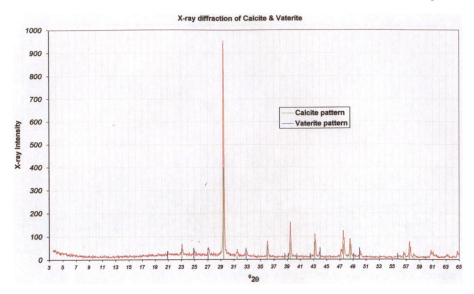

Figure 3 *XRD Diffraction pattern of forms of calcium carbonate*

materials. Sensitivity of the method depends on the nature of the material and ranges between 1 and 25% by weight, an average value being around 5%.

XRD is used for the identification of crystalline materials such as pigments, metal powders, organic materials and salts. Non-crystalline materials lacking a regular crystal lattice, such as glass, do not produce a clear pattern.

Infrared Spectroscopy. Infrared radiation is found between the higher energy (and shorter wavelength) visible light and the lower energy (and longer wavelength) of radio waves. When infrared light falls upon a compound it will interact with the outer shell electrons of a molecule, the ones that bond the atoms together to form that molecule. Thus, the technique can identify different types of bonding arrangements between atoms. Since bonds are not fixed in space but vibrate by stretching (distance between the atoms changes) or bending (the angle formed by the bond between the atoms varies), each type of movement will absorb a specific energy. The infrared absorption spectra produced can be divided into two regions, the group frequency region and the fingerprint region.

The group frequency region falls approximately between 4000 to $1400\,\mathrm{cm}^{-1}$, and the absorption bands in it may be assigned to vibration of pairs of two (or sometimes three) atoms. The frequency is characteristic of the masses of the atoms involved and the nature of their bond, ignoring the rest of the molecule. Therefore, IR spectra are useful for determining the presence of functional groups in organic compounds: alcohols (—OH), ketones (=CO), amines

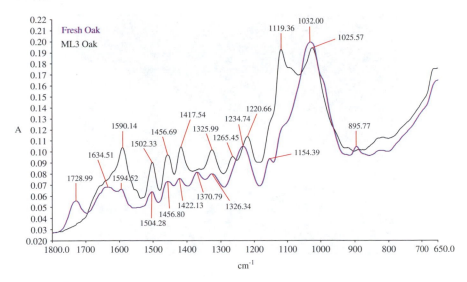

Figure 4 *FTIR spectrum of fresh oak and ancient oak (ML3)*

(—NH$_2$), *etc*. As a result, correlation charts tabulating the vibrational frequencies or absorption bands of the various functional groups are used.

In the fingerprint region (approximately 400–1400 cm^{-1}) the absorption bands correspond to the vibrations of the molecule as a whole, and are therefore characteristic for each molecule, and can serve as unambiguous identification by comparison with a standard. However, long polymers of a given family, such as long chains of fatty acids can give almost identical spectra.

The technology in infrared analysis has been improved significantly by the use of FTIR, which has increased the sensitivity of this method significantly (see Figure 4).

3.2 Interaction of Elemental Particles with Matter

Materials are made up of chemical compounds that in turn are formed by atoms. Atoms have essentially a dense nucleus around which a cloud of electrons balances the electrical charge of the nucleus. If high-energy electrons are shot at matter, some will bounce off the surface, some will penetrate into it producing X-rays, while others will interact with the higher energy electrons (inner shell electrons) of the atoms. These are the principles used for SEM, EDS and wavelength dispersive X-ray spectroscopy (WDS).

Scanning Electron Microscopy. A useful tool for investigation of the surface appearance of materials and any changes after treatment is the SEM. This tool is widely available in universities and some art museums. The principle of

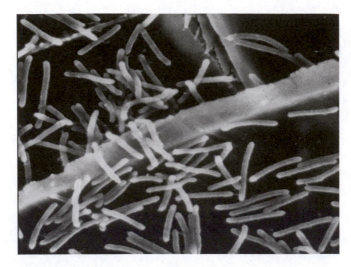

Figure 5 *SEM photomicrograph of wood surface colonised by bacteria*

operation is to generate an electron beam, with a range of 1–30 kV, at the top of a high-vacuum column. An electro-magnetic lens focuses the beam into a fine-spot size that then impinges onto the surface of the object to be examined. The unique feature of the SEM is that it has a set of scanning coils that drives the beam across the surface of the object, left to right, in a repeating fashion, similar to reading a book. As the beam interacts with the surface of the object different signals are generated from the sample surface. These surface interactions are collected by appropriate detectors, as the beam scans the surface, and are amplified and displayed on monitors (for a very readable introduction to SEM, see Postek *et al.*, 1980).

The main image generated by SEM is a surface image, generated from low-energy electrons (called secondary electrons), that are knocked out of the surface by the primary, high energy, beam. A black and white, 3-dimensional image of the surface is generated and presented on a cathode ray tube or computer monitor. Analysis of the surface morphology can produce much useful information (see Figure 5).

Another useful method that the SEM provides is that of viewing the surface of the sample with backscattered electrons. These are higher energy electrons from the primary beam that have been "bounced back" or scattered from the surface of the sample. The amount of back-scattered electrons depends on the atomic or molecular weight of the sample. The higher the atomic weight the higher the quantity of back-scattered electrons and the greater the contrast in image generated among elements of different atomic weight. Information collected in this mode can generate elemental distribution maps of the sample surface. Figure 6 compares two SEM photomicrographs of a corroded

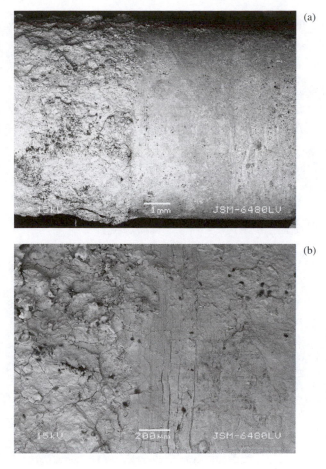

(a)

(b)

Figure 6 *Scanning electron micrograph of a corroded iron bolt (a) SEI and*
(b) back-scattered electron image

iron bolt; one image produced by secondary electrons (SEI) and the other by
back-scattered electrons.

Energy Dispersive X-ray Spectroscopy and Wavelength Dispersive Spec-
troscopy. Another useful signal collected from interactions of the primary
beam with the sample surface are the X-rays generated within the sample.
These can be analyzed both as a function of their energy (EDS) or their wavelength (WDS). EDS collects a range of energy signals, typically in the
0–10 KeV or 0–20 KeV range with a multi-channel SiLi detector. Each element in the sample produces a characteristic energy that can be collected and
displayed on a monitor (see Figure 7).

EDS is a fast technique that can produce qualitative elemental information within a few minutes. Although the computer will process results in a

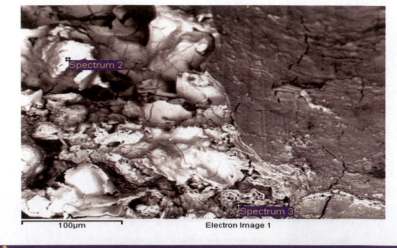

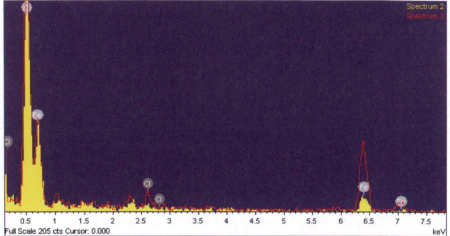

Figure 7 *SEM and EDS of an iron bar*

quantitative manner, care must be taken in its interpretation, as it may be indicative but not definitive. Some samples, *e.g.* polished metals, may yield quantitative results whilst powdery samples will be indicative of relative concentration at best. The sensitivity of this technique is typically in the 0.1–0.5% range, depending upon element and matrix.

WDS information is collected with a specific crystal detector that is rotated through the range where each specific element's diffraction wavelength is found. Different crystals are also used for different wavelength ranges and the geometry of the primary beam, sample and crystal are important for proper detection and collection of the information. WDS is much slower than EDS, but is generally one order of magnitude more sensitive than EDS. As a result, both techniques can nicely complement each other, as they have different

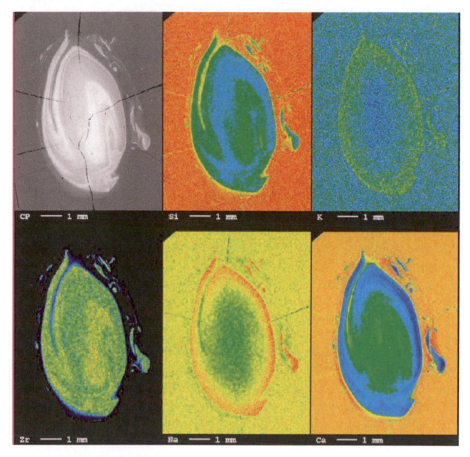

Figure 8 *SEM and dot mapping of inclusions in glass*

strengths and weaknesses, to produce a more accurate composition of the sample. Elemental maps may also be generated from EDS or WDS data. Figure 8 compares dot mapping with regular SEM images for inclusions in glass.

3.3 Dating Methods

The dating of an object is critical both from an archaeological point of view and from a museum point of view. In the first case, it will allow a date to be attributed to a settlement, or provide a chronological sequence in an excavation. In the second case, it may confirm or refute the assigned provenance of the object. Unfortunately, there is no single dating technique applicable to all materials. For example, obsidian objects can be dated by the thickness of the hydration layer that forms on this volcanic glass once an object has been fashioned out of it. Wood can often be dated by dendrochronology, which counts the number of rings that trees develop with growth. Apart from the radiocarbon dating discussed below, there are several radioactive methods for dating other elements (see Dijkstra and Mosk, 1981). Each of the available dating techniques is limited to only some types of materials, and all of them are complex techniques that require much experience to be able to assess the data produced.

Radiocarbon 14 dating.   Radiocarbon dating uses the physical and biological phenomenon of incorporation of the radioactive ^{14}C isotope into the biosphere (all living plants and animals; see Bowman, 1990). This isotope is continually formed in the upper atmosphere where it is rapidly converted into "marked" carbon dioxide $^{14}CO_2$. This in turn mixes within the atmosphere with the "normal" carbon dioxide (mostly $^{12}CO_2$) and in waters and is eventually utilized by living organisms and incorporated into their tissue. The ratio between these two isotopes is about $1:10^{12}$. The activity associated with the ^{14}C in this equilibrium is about 15 disintegrations/minute.

Once an animal or plant dies, it ceases to accumulate CO_2 and the level of ^{14}C slowly decays as it disintegrates by ß-emission. It takes approximately 5600 years for the number of ^{14}C atoms to decay by one half; this is called the half-life of the isotope. This rate of disintegration is directly proportional to the number of radioactive atoms present. Since the decay constant for ^{14}C is known, by measuring the decay rate of the object at the present time (and knowing the original decay rate) it is possible to calculate the age of the object.

Dateable materials are the organic matters that includes pollen, bone, charcoal, shell and many animal products and remains. There are a number of assumptions that go into dating and are best left to experts in the field to take into account when trying to determine a ^{14}C date for any material.

Thermoluminescent dating (TL).   This technique is useful for dating pottery and ceramics (Wagner *et al.*, 1983). It is based on the cumulative effect of radiation from disintegrating radioactive isotopes present in minerals of most rocks, clays and soils. The ionizing radiation may cause electrons to detach from their parent atom and become trapped in lattice defects of the material. As the material ages, defects accumulate. When a ceramic object is fired, trapped electrons are freed and, since radioactive minerals are incorporated in the object, defects start to accumulate again.

TL uses this same principle to date the object. A small powdered sample of the object is heated and by measuring the emission of light in excess of the incandescent glow that is produced the date of the firing can be calculated. The technique requires calibration of each particular sample, since every clay particle will accumulate defects at a different rate. Although the operational principle of the method appears simple, it requires experience to interpret the data correctly.

4   DIAGNOSIS OF DETERIORATION PROCESSES

During the examination of the object it is important to assess its condition and determine whether it has deteriorated over time. If so, what were the causes that led to this change and is it still occurring?

For museum objects, in many cases the deterioration occurred prior to its acquisition by the collection; however, some deterioration agents may have entered into the body of the object, become dormant and continue to be a problem. Such is the case with archaeological ceramics that have been contaminated with soluble salts, or of wooden objects that bring with them insect infestation. Although in some cases it may appear that the deterioration is not progressing, this may re-appear at any moment prompted by a change in environmental conditions (temperature and relative humidity (RH)). The detection of these possible dormant deterioration factors is critical since it will allow a conservation treatment to be designed specifically to address them. For example, some potassium glasses are susceptible to changes in RH, around the 40% level. If the RH varies here, cracking and crizzling may occur. Some modern materials, *e.g.* Masonite®, absorb moisture very rapidly, but release it slowly. This means that transient spikes in humidity may lead to moisture retention by a material, with subsequent fungal growth.

5   CONSERVATION TREATMENTS

Conservation treatments can be divided broadly into four main categories: cleaning, desalination, consolidation and disinfestations. Within each of these categories there can be several sub-categories for the different cases and different materials that have to be addressed, such as the de-acidification of paper. The following is an attempt to give a broad picture of the aim of these treatments and the problems that need to be considered. Three principles should guide any conservation intervention:

- Reversibility or retreatability
- Compatibility
- Minimum intervention

The concept of a "reversible" treatment was suggested by the traditional periodic removal and re-application of varnishes to oil paintings. However, this cannot really be extended to all objects. For example, an archaeological object may have an incrustation resulting from its long burial. Removal of this incrustation will result in the loss of information about the type of incrustation and the mechanism by which it was formed; information that may be crucial from an archaeological or historical point of view. Furthermore, this process is not reversible. The decision to remove the incrustation is based upon a judgment of values, *e.g.* the value of the object itself is more important than its history; hence, part of the historical value is sacrificed as a function of the aesthetic value of the object.

A further example is furnished by the case where a fragile object requires consolidation. How can one expect that this treatment will be "reversible", *i.e.* that it can be completely removed without risk of stressing the object even more by this removal? These considerations have led to replacing the "reversibility" concept with that of "retreatability" (see Teutonico *et al.*, 1997). In this case, any treatment applied should not preclude or hinder any future treatment. An example of a totally irreversible treatment is the complete impregnation of stone statues with *in-situ* polymerized methyl methacrylate.

The concept of compatibility asserts that any material applied to an object should be compatible with it, *i.e.* its properties should be as close as possible to eliminate future stresses. The problems that appear when different materials are put together in an object are illustrated by many of the museum objects that present conservation problems today. Consider, for example, an object that has a metal inlay in wood, as compared to stone inlay in stone. The latter has far fewer problems than the former, precisely because their properties, such as expansion coefficients upon heating or during changes in relative humidity, are similar.

Finally, the minimum intervention concept is the most important. This is because, as discussed above, any action taken on an object is irreversible, as time is irreversible. Any treatment applied changes the object and will interfere or preclude other analyses that one might want to do in the future. This is particularly important given the fast development of more sophisticated analytical techniques that can provide data that had not even been envisaged only 50 years ago.

5.1   Cleaning

This is probably the most frequent treatment performed on museum objects. It can range from removal of loosely-deposited surface dust to that of hard, adherent concretions. In general, it is considered as a simple and self-evident task and consequently not much thought is given to it. However, it is not a simple task and, in many cases, difficult decisions have to be made about how it has to be performed. If a metal object has a heavy calcareous incrustation covering most of it, would removal of this incrustation, which may result in the removal of the original patina the object had, be acceptable? Loss of the original patina means that part of the information that patina carried (How was that patina formed? Was it natural? Was it the result of an intentional treatment?) would be lost, and with it, part of the authenticity of the object.

If the case presented by a metal object is difficult, it is even more difficult in the case of a stone object, because the concept of "patina" is yet to be generally accepted for this material. In some cases, for example some porous limestones, a hardening of the surface will result from their exposure to an unpolluted environment, referred to as "calcin" by the French. Some ferruginous sandstones may develop a thin, black surface layer (not to be mistaken with a black gypsum crust) when exposed outdoors, through migration of iron and manganese oxides to the surface. These can be considered "natural" patinas equivalent to those formed by metals such as aluminum and copper. On the other hand, there are man-applied patinas, such as the "scialbature" applied to marble sculptures and, in some cases, corresponding with the formation of an oxalate surface deposit.

At heart of the matter is to what degree should the object be cleaned? Does it have to look new? Or should it show its age? The current school of thought is that it should show its age. But this was not always the case. A good example is provided by the Elgin marbles – the marble sculptures acquired by Lord Elgin from the Parthenon in 1801–1802. These sculptures still had traces of polychromy when they were acquired; however, at the time, the perceived "classical" marble sculpture was white. Hence, the sculptures were cleaned down to the bare marble (Oddy, 2002). This clearly emphasizes the point that conservation is a cultural operation carried out by technical means. The conservator has to use his judgment in assessing the degree of cleaning required and this needs to be confirmed in conjunction with the historians and art historians.

Cleaning techniques may range from a simple dusting with a soft brush, to careful removal of light deposits with aid of a scalpel, or in some instances using spit on a cotton swap to effectively remove the deposit, with the enzymes in the spit providing a cleaning action. For harder deposits, microabrasive techniques can be used, or, in some cases, poultices may be applied.

5.2 Desalination

If the object contains potentially damaging soluble salts, as is the case for most archaeological ceramics, the approach to their removal will depend on the amount of salt present in the object. If it contains only a relatively low amount of them, then brushing off (or vacuuming off) the efflorescence may be the most effective method. However, if the object contains a large amount of them, then either poulticing or successive baths in distilled or de-ionized water may have to be the procedure used. These approaches need extra care, since successive poulticing may affect the surface finish on the object and, in the case of desalination by water immersion, the thoroughly water-soaked material will be far more delicate to handle.

5.3 Consolidation

When an object turns fragile because of its age and the consequent deterioration of the material, as is the case of textiles, papers and many objects from natural history collections; or, because of the deterioration suffered by insect infestation or the presence of soluble salts, the object needs to be consolidated.

There are many products and procedures for carrying out a consolidation treatment, but they all involve introducing another material into the object. As discussed above, the ideal "reversible" treatment does not exist. Furthermore, once the treatment is introduced it may interfere with future objectives. For example, previously buried bones in natural history collections can be very friable and have been consolidated systematically to allow their presentation. This, however, precludes that they can be dated after treatment (especially if organic resins have been used), since more carbon has been introduced in the sample. It may also, preclude analysis of DNA, or other biological molecules.

5.4 Disinfestation

The problem of insect and microbial infestation of art is centuries old. Techniques to control it have included such things as herbal treatments, fire smoke and, most recently, chemicals. All have provided some degree of effectiveness if not complete control of the pests, but too often the treatment, whilst meant to save the art, has created damage of its own. A recent approach developed at the Metropolitan Museum of Art, and other institutions, is a non-chemical treatment using oxygen-free environments. The best choice for generating an oxygen-free environment is to use argon gas. This gas is heavier than oxygen so that it sinks to the bottom of the enclosure or bag forcing oxygen out of the object and pushing it to the top of the bag. This approach has proven very successful for insect control and also kills some species of fungi, if the residual oxygen level is low enough (but has no effect on spores).

The concept of anoxic treatment is simple, and is the same for any anoxic gas used. It consists, essentially, of the following three steps:

1. Isolate the object from the oxygen-rich environment;
2. Replace the oxygen-rich air with an anoxic (oxygen-less) air to the desired residual oxygen level; and
3. Wait until the insects die and then remove the object from its anoxic environment.

Whilst simple in concept, each step requires an understanding of environmental, physical and biological factors that may affect the procedure. Perhaps the most important of the three steps is isolation of the objects. Isolating an object requires construction of a suitable barrier around the object. This means any enclosure system must successfully maintain such a low level of oxygen for extended periods of time, ideally with a minimum of intervention and cost. The enclosure may be hard-walled or soft-walled. The soft-walled, *e.g.* heat-sealable oxygen barrier film, gives the flexibility to create any sized enclosure specific to the need and is very portable – it permits an enclosure to be constructed on-site, reducing potential damage and costs of shipping.

How low an oxygen environment and for how long must it be held? To determine how long insect-infested objects must remain in a given argon environment has been determined by actual measurement of insect respiration before and after treatment by use of a gas cell FTIR system to measure the CO_2 produced by insects or fungi. Using this system it is possible to detect the presence of one insect in a 10-L bag within 4 h. A treatment length of 3–4 weeks, at less than 500 ppm (0.05%) oxygen in argon, is the recommended time (depending on insect species, life stage, size of object, and temperature and humidity conditions).

6   PREVENTIVE CONSERVATION

Following the considerations of the problems presented by conservation treatments, the ideal would be that these should not be necessary or are only required in a minimum of cases. From this consideration the idea of "preventive conservation", that is, prevent the need of a conservation treatment, was born.

The concept of preventive conservation is to try and maintain, and monitor, the most appropriate environmental conditions around the art so as to reduce stresses on the objects and preserve them with no, or minimum treatment. For insect control, this is referred to as integrated pest management (IPM), but the principals of IPM can be extended to control of other problems in storage of art. IPM is basically good house-keeping: keep moisture away from materials – whether in the form of liquid water or high humidity; keep food, plant and animal products isolated from the art; keep an airflow around the objects to reduce the risk of moisture buildup; keep the environment clean. All objects are not the same and some may require special storage or display conditions, for example, a potassium-rich glass may "crizzle" if the humidity oscillates above and below 40% RH, whereas a soda-rich glass may not show any response to the change.

7   CONCLUSIONS

It is clear, from the brief discussion of but a few of the methods used in conservation that this is a large, interdisciplinary field. While the contribution of curators, art historians and scientists is fundamental, the conservator probably has the most difficult task: actually working on the object. This is an enormous responsibility and requires that the conservator acknowledges it as such. The key approach in conservation is respect for the object to be conserved. This requires that the conservator be a good observer, a talented craftsman and endowed with the clear discernment required to determine the exact point to which any procedure should be carried to.

What contributes to the complexity of the conservation field is the juxta-position of the rigor needed in the methods used for analysis to the varying approaches that conservation requires, depending on time and location. Thus, while analytical methods rely on the use of standards against which measure-ments can be compared, conservation definitely does not have such standards. Furthermore, conservation relies heavily on the crafts component – a point that is being forgotten as the application of science has gained terrain in conserva-tion. But science can never, nor should it, replace conservation. The contribu-tion of science, with regard to conservation, is to help understand deterioration mechanisms that affect works of art and other cultural objects. This will certainly aid in diminishing their deterioration rate and assist in devising better preventive conservation approaches and better conservation methods, but the applications of these methods will still depend on the sense and sensibility of the conservator.

REFERENCES AND FURTHER READING

M.W. Ainsworth, *Art and Autoradiography: Insights Into the Genesis of Paintings by Rembrandt, Van Dyck, and Vermeer*, The Metropolitan Museum of Art, New York, 1982.

S. Bowman, *Interpreting the Past: Radiocarbon Dating*, British Museum Publications Ltd, London, 1990.

P.A.T.I. Burman, "Hallowed antiquity": Ethical considerations in the selec-tion of conservation treatments, in: N.S. Baer and R. Snethlage, (eds), Saving Our Architectural Heritage: The Conservation of Historic Stone Structures, Dahlem Workshop Report ES20, Chichester, Wiley, New York, 1997, 269–290.

T. Carunchio, Dal restauro alla conservazione, Introduzione ai temi della conservazione del patrimonio architettonico, Edizione Kappa, Roma, 1996.

G. Dijkstra and J. Mosk, The analysis of art through the art of analysis, *Trends Anal. Chem.*, 1981, **1**(2), 40–44.

M. Ferretti, *Scientific Investigations of Works of Art*, ICCROM, Rome, 1993.

F.M.A. Henriques, Algumas reflexões sobre a conservação do património edificado em Portugal, in 2 ENCORE, *Proceedings of the Encontro sobre Conservação*, Reabilitação de Edifícios, Laboratorio Nacional de Engenharia Civil, Lisboa, 1994, 67–86.

R.J. Koestler, Insect eradication using controlled atmospheres and FTIR measurement for insect activity, *ICOM Committee for Conservation 10th Triennial Meeting*, Washington, DC, 1993, 882–885.

R.J. Koestler, Detecting and controlling insect infestation in fine art, in C.M. Stevenson, G. Lee and F.J. Morin (eds), Pacific 2000, *Proceedings of the 5th International Conference on Easter Island and the Pacific*, Easter Island Foundation, Los Osos, CA, 2001, 541–545.

R.J. Koestler, P. Brimblecombe, D. Camuffo, W. Ginell, T. Graedel, P. Leavengood, J. Petushkova, M. Steiger, C. Urzi, V. Verges-Belmin and T. Warscheid, How do environmental factors accelerate change? in W.E. Krumbein, P. Brimblecombe, D.E. Cosgrove, and S. Staniforth, (eds), *The Science, Responsibility, and Cost of Sustaining Cultural Heritage*, Wiley, New York, 1994, 149–163.

R.J. Koestler, S. Sardjono and D.L. Koestler, Detection of insect infestation in museum objects by carbon dioxide measurement using FTIR, *Int. Biodeter. Biodegr.*, 2000, **46**, 285–292.

J.S. Mills and R. White, *The Organic Chemistry of Museum Objects*, Butterworths, London, 17–22, 1987.

A. Oddy, The conservation of marble sculptures in the British museum before 1975, *Stud. Conserv.*, 2002, **47**, 145–154.

P.L. Parrini, (ed), Scientific Methodologies Applied to Works of Art, *Proceedings of the symposium*, Florence, May 2–5, 1984, Montedison Progetto Cultural, Milan, 1986.

M.T. Postek, K.S. Howard, A.M. Johnson and K.L. McMichael, Scanning electron microscopy a students handbook, *Ladd Res. Ind. Inc.*, XV, 1980, 305pp.

M. Realini and L. Toniolo (eds), The Oxalate Films in the Conservation of Works of Art, Editeam, Castello d'Argile (BO), 1996.

P.M.S. Romão, A.M. Alarcão and C.A.N. Viana, Human saliva as a cleaning agent for dirty surfaces. *Stud. Conserv.*, 1990, **35**, 153–155.

A. Sinclair, S. Slater and J. Gowlett (eds), *Archaeological Sciences 1995. Proceedings of a conference on the application of scientific techniques to the study of archaeology*, Oxbow Monograph 64, Oxbow Books, Oxford, 1997.

C. Tavzes, F. Pohleven and R.J. Koestler, Effect of anoxic conditions on wood-decay fungi treated with argon or nitrogen. *Int. Biodeter. Biodegra.*, 2001, **47**, 225–231.

J.M. Teutonico (rapporteur), A.E. Charola, E. De Witte, G. Grassegger, R.J. Koestler, M. Laurenzi Tabasso, H.R. Sasse and R. Snethlage, Group Report: How can we ensure the responsible and effective use of treatments (cleaning, consolidation, protection)? in N.S. Baer and R. Snethlage (eds), *Saving Our Architectural Heritage: The Conservation of Historic Stone Structures*, Dahlem Workshop Report ES20, Chichester, Wiley, New York, 293–313, 1997.

G.A. Wagner, M.J. Aitken and V. Mejdahl, *Handbooks for Archaeologists: No. 1. Thermoluminescence Dating*, European Science Foundation, Strasbourg, 1983.

R. Wihr, 15 Jahre Erfahrung mit der Acrylharzvolltränkung (ATV), Arbeitsblätter für Restauratoren, 1995, **28**, (no. 1, Gruppe 6), 323–332.

CHAPTER 3

Paper

VINCENT DANIELS

Conservation Department, Royal College of Art, Kensington Gore,
London SW7 2EU, UK

1 THE CONSTITUENTS OF PAPER

1.1 Fibres

Everyone can recognise a piece of paper when they see one, but it is more difficult to *define* what paper is. Although paper can be made from synthetic fibres,
the vast majority of paper is, and had been, made from cellulose fibres. The
pages of this book contain cellulose fibres derived from wood. Plants contain
fibres made of a carbohydrate polymer called cellulose. Not all parts of plants
contain cellulose but it still makes up about a third of the mass of the higher
plants and is the most abundant biopolymer on earth. Both starch and cellulose
may be thought of as polymers of glucose, however, the cellulose molecule is
different from that of starch because the glucose molecules (strictly speaking
anhydroglucose) are linked together in a different way, and the repeating unit
in cellulose is the dimer cellobiose. A cellulose molecule is shown in Figure 1.
The link between the rings is called a glycosidic linkage.

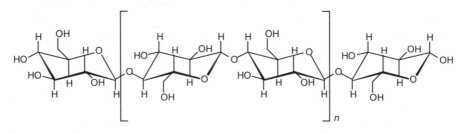

Figure 1 *Structure of cellulose*

When paper was first made in Medieval Europe in about 1200 AD, the best source of easily separated cellulose fibres was rags made of cotton or linen. Cotton occurs naturally as nearly pure cellulose in the fluffy seed heads of the cotton plant and can be used for textile- or paper-making with little preparation. Linen fibres occur in the stems of the linen plant but the fibres have to be freed from the stem matrix. Linen fibres are separated by a combination of microbiological, chemical and mechanical methods by first softening the stems of the plant by leaving them in water for several weeks (retting) and then subjecting them to beating and further purification. Nowadays, little rag paper is produced, but some is used by artists and calligraphers. Modern papers are mostly made from wood cellulose fibres that are difficult to purify and chemical processes have to be used which can cause significant damage to the cellulose fibres/molecules while effecting the separation.

Traditional Chinese and Japanese papers are made from cellulose fibres derived from locally-occurring plants. The fibres are separated by processes similar to those used on linen. Long fibres with relatively undamaged cellulose produce strong papers with good permanence. The first papers ever made are believed to have been made in China in 105 AD.

Paper is made from a suspension of fibres in water. The suspension is drained through a fine mesh so the water drains away and the fibres are retained to form a mat on the mesh. On drying, the fibres stick together and form a sheet of paper. Many paper experts define paper as a material made by a process that involves the separation of fibres into a suspension with subsequent formation into a sheet.

It can be seen from the formula for cellulose that there are many hydroxyl groups all along the chain. These –OH (hydroxyl) groups are attracted to one another by the attraction of a partial negative charge on the oxygen atom for the partial positive charge on the hydrogen atom (hydrogen bonding); this is the principal mechanism explaining why paper fibres are held together in a sheet of paper. Another mechanism is the tangling together of the long, thin fibres.

An important step in papermaking is the modification of the paper fibres by "beating". When undamaged, paper fibres are long and thin with relatively little surface roughness. If damaged in the intended way, the ends and surfaces of the fibres can be induced to fray into thinner subunits, a mental picture may be of the way our hands have fingers at the ends. This process is called beating because that is exactly how the process is traditionally done, by hitting the fibres with big mallets. Beaten fibres have a much bigger area for the formation of hydrogen bonds and generally form stronger sheets of paper. Industrially, beating is performed in a machine in which the fibres pass between two moving metal blades that subject the fibres to great stress and shear the fibres apart.

Handmade paper is formed by dipping a tray with a mesh base (the mould) into a suspension of fibres in water, lifting the mould from the vat of fibre suspension, and then allowing the water to drain away. The paper is then removed from the mould and allowed to dry. Machine-made paper is made by allowing the fibre suspension to flow onto a moving mesh (the web), thus forming a continuous strip of paper which may be dried and pressed on the machine before being made into a roll for storage. One of the arts of paper-making is to form the suspension of fibres in the first place and then stop the fibres becoming entangled before they settle on the web. When clumps of fibres occur in paper, they are called knots. Modern papermakers use synthetic suspending agents to avoid knots, but Oriental artisans used mucilage made from seaweed.

Papermaking fibres usually contain substances other than cellulose. There are often carbohydrate polymers other than those that are polymers of glucose. There are many sugars present in plants, which may polymerise to form carbohydrate polymers and which become included in plant cells from which fibres are obtained. Carbohydrates similar to cellulose, which are made out of a mixture of different sugars, are called hemicelluloses. Often the hemicellulose is named after the principal sugar that they contain, *e.g.* xylan, a hemicellulose in wood, is named after the sugar xylose. The polymers may be branched, unlike cellulose molecules. With all their branching and mixture of different sugar monomers, hemicellulose molecules cannot pack closely together and do not contain ordered regions. Cellulose is a comparatively linear polymer and by hydrogen bonding will form regularly packed, ordered regions that have crystalline properties, *e.g.* they will diffract X-rays. Carbohydrates in solids with crystalline areas are relatively resistant to chemical agents trying to attack them because the attacking molecules find access difficult. Hemicelluloses are more reactive than cellulose because of the reactivity of the sugars they contain and because they have no crystallinity. Hemicelluloses can be removed from impure cellulose fibres by treatment with alkalis.

Some papermaking fibres, especially those derived from wood, contain lignin which is a large three-dimensional polymer made of building blocks containing rings of six carbon atoms called phenyl-propanes. The structures of the phenyl-propanes in grasses and in coniferous and deciduous woods are slightly different from one another, however, Figure 2 indicates how lignin is typically constructed.

Lignin is the stiffening material that gives wood its characteristic mechanical properties; it stops plants from falling over! Wood cellulose is so intimately associated with lignin in natural products that the vigorous physical and chemical treatments needed to ensure complete removal of lignin damage the cellulose.

Wood can be ground to a powder and then with a little further processing can be used for making a type of paper. These papers have low strength as

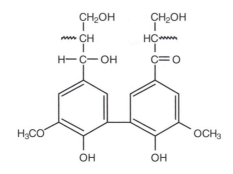

Figure 2 *The basic structure of lignin*

there are few free fibres and the opportunities for hydrogen bonding between the impure cellulose fibres is limited. These "groundwood" or "mechanical wood pulp" papers are cheap to make and are used in paperback books and newspapers. Changes in production methods for the cheapest papers mean that many of those papers are now more permanent.

It is now possible to consider two types of cellulose-based writing surface that do not come into our definition of paper but which are important to the paper conservator. Although paper was first made in China in about 105 AD, the Ancient Egyptians made a writing surface of what we call "papyrus" from about 3000 BC. The papyrus plant (*Cyperus papyrus*) has tall stems, approximately triangular in cross-section, inside which there is a high concentration of cellulose fibres as part of a cream-coloured, spongy material. Once the tough, green outside layer is removed, thin strips of papyrus, a few centimetres wide, are cut and laid side by side to form the shape of a small sheet. More strips are then laid at right angles on top of these and the sheet is then pressed and dried. Although this material contains fibres of cellulose, they are not separated from one another, so it is not strictly paper. Another material, which is similar to, but not paper, is barkcloth also known as *kapa* or *tapa* in the Pacific islands where much of this material is made. Barkcloth is made from the cellulose fibres that are found beneath the bark of certain trees and shrubs (inner bark). The fibres are first retted and then, beaten with grooved mallets to further soften them and form sheets of matted fibres. While still wet, the sheets are folded up on themselves and beaten further to spread and reform the sheets. Barkcloth can be made into surfaces for writing on, made into clothing or used in a multitude of ways.

1.2 Paper Size

The type of filter paper found in the chemical laboratory is made from cellulose fibres only, but is of little use as a writing or printing paper as it is too

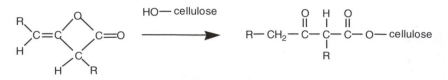

Figure 3 *The reaction of alkyl ketene dimer with cellulose*

absorbent for inks and has little wet strength. Generally, paper is sized to make it more hydrophobic (water repellent). Sizing also changes the surface texture and both the wet- and dry-strengths of the paper. Traditionally, a hot solution of gelatine was brushed onto the paper in its unsized state (water leaf) called tub sizing. Papermaker's alum (aluminium sulfate) was sometimes added to the gelatine. Chemists will realise that this is a misnomer as this chemical is not a true alum, however, papermakers consider its properties similar to aluminium potassium sulfate, which is an alum and also used by papermakers. Papermakers say that alum is added to "harden the size" but it is possibly of much value as a biocide to stop the stored gelatine solution from developing fungal colonies while it is being stored cold. A popular sizing agent from the mid-19th to the mid-20th century was alum/rosin size. Rosin is a natural product derived from coniferous trees and is chemically a mixture of rosin acids. It can be used in conjunction with alum to produce a size that produces optimum results on paper. In water, the trivalent, positively-charged aluminium ion is attracted to the negatively-charged fibre surface and there forms a link to attach the negatively-charged carboxylic acid group on the rosin acid. There are many ways in which rosin can be used as a paper size and several theories on how they work.

Modern papers are often sized with an alkyl ketene dimer that reacts with hydroxyl groups on the cellulose, and the large alkyl group imparts water-repellency (see Figure 3). The product is a white solid that is emulsified with water for application onto the paper.

2 THE DIRECTIONAL PROPERTIES OF PAPER

In most processes for the manufacture of paper, the fibres may align themselves in a preferred direction. In machine-made paper, the orientation is particularly marked, as the fibres are oriented in the direction in which the web is moving, the machine direction (MD). The MD of a piece of paper is of particular importance in determining the properties of a piece of paper. The direction across the paper at right angles to the MD is called the cross direction (CD). When paper is made from pulp, the smallest particles will penetrate furthest into the paper as the water drains through it. So for this and several other possible reasons, the paper also has differing properties in the thickness of the paper (ZD). The paper conservator will be aware of the different surface textures that occur on the different sides of paper.

A simple experiment can be used to determine MD. Take a sheet of paper from a newspaper or magazine and cut it into a square, keeping the sides of the square parallel to the original edges of the page. Bending the paper into a U-shape first along, then across the sheet will show that the paper bends more easily in one direction, the direction of the "tunnel" made by bending the paper is the MD. The paper also tears more easily in the MD. Gently bending a corner of a page of this book will probably show that the MD is parallel to the spine; this enables the pages to be turned more easily. There are some other MD-related properties that are related to the changes of paper when it interacts with water; these will be described later.

Papers come in different types; they often contain fillers that may be minerals or polymers. Papers may have surface coatings to provide a good surface for printing, to texture them or may have a layer of adhesive. Thick sheets of paper can be formed by casting a lot of fibres onto the web or laminating thinner papers together with adhesives to form cardboard. Some papers are dyed, waterproofed, fireproofed or perfumed. Paper is the support for photographs and can be made into books and magazines, or can be made into all manner of objects, including boxes. Paper with printing, drawing or painting on it and paper constructions are much more complicated than just a sheet of paper and any ageing and conservation process of a sheet of paper will be influenced by the media on the surface and *vice versa*. These examples are intended to convey the realisation that the subject of paper conservation is very wide-ranging and it is beyond the scope of a single chapter to attempt a comprehensive overview of all the scientific aspects of the work of the paper conservator.

3 HOW PAPER INTERACTS WITH WATER

The cellulose molecule is liberally supplied with hydroxyl groups, which form hydrogen bonds with both water molecules and hydroxyl groups on other cellulose molecules. Cellulose molecules prefer to bond more with other cellulose molecules than water, so cellulose fibres do not dissolve in water. In a fibre, some regions contain regularly ordered cellulose molecules that are in crystalline domains, while other areas are disordered and said to be amorphous. Natural cellulose fibres will give a good X-ray diffraction pattern, which can be used to study the crystallinity of fibres. In the crystalline areas, the molecules of cellulose are so closely packed that even small water molecules cannot gain access, however, the water will go into the amorphous areas and cause swelling and plasticise the fibre, making it more flexible.

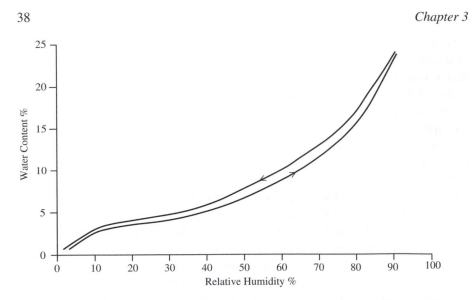

Figure 4 *How cellulose fibres absorb and desorb water with changing relative humidity; upper curve shows desorption and lower curve shows absorption*

Cellulose will reversibly absorb water vapour as the relative humidity (RH) is increased and decreased. Figure 4 shows a typical absorption/desorption curve for cellulose. Note that the two curves do not coincide, the desorption curve lags relative to the absorption curve; this is called hysteresis.

When water is absorbed, the dimensions of the fibres change. As the general orientation of the cellulose molecules is along the fibre, most of the expansion occurs sideways. The increase in dimensions is proportional to the water content, thus a graph of fibre diameter versus RH has the same shape as the graph in Figure 4 and hysteresis is again exhibited.

Individual fibres can expand in width up to 20% when going from 0 to 100% RH but lengthen perhaps about one-tenth of that percentage. The paper does not expand up to 20% because the effect is attenuated by the microstructure of the paper. Typical maximum expansion of paper is about 2–3%. As most fibres are aligned along the MD, the hygroexpansivity of paper (produced by RH changes) is mostly in the CD. This may be demonstrated on a piece of paper; by wetting one side of a small square, the paper will form a U-shape along the MD as the fibres on the wetted side expand sideways while those on the dry side do not.

Newly-made paper contains dried-in stress introduced during the drying of the paper. The stress will be removed if the paper goes to a sufficiently high RH during storage, when the paper may shrink irreversibly. Old papers do not appear to have dried-in stress, possibly this has been relieved during repeated small RH fluctuations over many years.

Uneven dimensional changes have serious consequences for the paper conservator. If a piece of paper expands on one edge while an adjacent area remains unchanged, then there will be a connecting area where there is stress. In areas of sufficiently high stress, the paper may undergo irreversible deformation with resulting loss of flatness. The production of such resulting wavy surfaces is called "cockling". The phenomenon is often seen on damp-damaged books where water has gained local access and the paper cannot expand freely as it is confined by other pages and the binding. Aqueous conservation treatments are difficult to perform on bound books because of the dimensional changes in the paper. The paper conservator has only to join two pieces of paper together to generate the potential for cockling as the two pieces will have differing hygroexpansivities and one or both may have dried-in-stress. The conservator may take care to use similar papers that have been stored for some time, align the MDs and avoid getting the paper too wet if using a water-based adhesive.

4 DETERIORATION OF PAPER

Paper objects may suffer from damage by vermin, insect and microbial attack, theft, fire and flood. However, from a chemist's point of view there are two important mechanisms by which paper deteriorates, both of which may change its hue and make it weaker; these mechanisms are acid-catalysed hydrolysis and oxidation. Papers vary a lot in their stability and the worst can become so brittle that they become unusable. Papers made from rags using traditional methods are usually extremely stable, but old papers made from poorly processed wood pulp are often very weak.

4.1 Acid-Catalysed Hydrolysis of Cellulose

The first comprehensive and systematic work on the reasons for deterioration of paper was performed by W.J. Barrow in the 1960s who firmly established the link between the acidity of paper and its permanence. Although we talk about the acidity of a sheet of dry paper, the acidity of paper, expressed as pH is only measured in an aqueous extract, generally produced by macerating 1 g of paper in 50 mL of pure water.

The pH is defined as $pH = -\log_{10}[H^+]$ where $[H^+]$ is the hydrogen ion concentration. On this scale, pH 7 is neutral, pH 1 is extremely acid and pH 14 extremely alkaline. Each pH decrease of 1.0 represents a 10 times increase in hydrogen ion concentration. For example, there is 10 times more active acid at pH 4 than pH 5.

Barrow measured the pH of many old pieces of paper and also measured their strength and found that alkaline papers were generally strong but acid papers were generally weak. There are many ways of measuring paper strength but Barrow used an instrument called the MIT folding endurance tester (see below for details of this test).

Barrow's other approach to the problem was to obtain several pieces of paper and age them artificially and follow the decrease in strength. The results reinforced the relationship between initial acidity and loss of strength on ageing. There is more about artificial ageing later in this chapter.

The loss of strength of paper produced by acidity is produced by acid-catalysed hydrolysis of the cellulose chains. In ambient RH, cellulose contains several percent of water that may participate in hydrolysis reactions. The point of scission is the link between the glucose (pyranose) rings. In paper-conservation science, the degree of polymerisation (DP), *i.e.* the average number of monomer units, is commonly used to discuss the molecular weight. Loss of DP does not produce an effect on the strength immediately; it has been proposed that the DP has to reach a critical point before loss of strength becomes apparent. Celluloses of an initially high DP, such as natural cellulose from cotton and linen, thus have an advantage over other types of an initially lower DP such as wood celluloses after extraction from wood. Loss of strength occurs most in amorphous areas as water and acids have difficulty in obtaining access to crystalline areas. Paper can start life acidic because of its method of manufacture and ingredients, or it may absorb acidic air pollutants such as sulfur dioxide and nitrogen oxides. Sulfur dioxide is believed to form sulfuric acid in the paper. Papermaker's alum produces an acid solution as it can be thought of as a salt of a weak base (aluminium hydroxide) and a strong acid (sulfuric acid). Rosin, used in some types of size, also contains rosin acids.

Alkaline degradation is possible but occurs much less readily than acid degradation. Degradation in alkaline conditions is of interest to the paper scientist mostly because of the degradation which may occur during analytical procedures rather than what may happen during the ageing of paper.

4.2   Oxidation of Cellulose

Paper exists in air that contains about 20% of oxygen. Cellulose is capable of reacting slowly with oxygen in ambient conditions (autoxidation). Such oxidation reactions involve intermediates called free radicals. In a simple compound A–B, the single bond between A and B contains two shared electrons, if both go to B then two ions are produced A^+ and B^-. If the electrons are shared between A and B, two free radicals $A^•$ and $B^•$ are produced because each has an unpaired electron. Such free radicals can participate in chain reactions in which the reaction of a free radical with a neutral molecule creates a product and another free radical (propagation). The process of oxidation starts when a free radical reacts with oxygen to make a peroxy radical ($ROO^{\bullet}$) that can then react further. Some autoxidation reactions could include:

$$RH + O_2 \rightarrow R^{\bullet} + HOO^{\bullet} \qquad \text{(initiation)}$$

$$R^{\bullet} + O_2 \rightarrow ROO^{\bullet} \qquad \text{(propagation)}$$

$$ROO^{\bullet} + RH \rightarrow ROOH + R^{\bullet} \qquad \text{(propagation)}$$

RH may represent any organic compound.

Autoxidation is slow in pure cellulose and it may be considered to take place at a negligible rate, however, certain conditions will initiate and accelerate free radical reactions. If light is absorbed by cellulose or a material associated with cellulose, *e.g.* a dye or lignin, the absorbing molecule's energy increases greatly. Many things may happen to this energy but one of them is that the energy can be transferred to another molecule, *e.g.* if lignin transfers its energy to cellulose, this is called photosensitisation of the cellulose by the lignin. The energetic cellulose could split open one of its chemical bonds producing two free radicals or the energy may be transferred to an oxygen atom which, then being in a highly energetic state (*e.g.* singlet oxygen), can oxidise the cellulose.

Metal ions that are capable of changing their valency, in particular, ions of copper and iron, are powerful oxidation catalysts as they can initiate oxidation reactions. During a valency change, negatively-charged electrons (e^{-}) are donated or accepted.

$$Cu^{+} \rightleftharpoons Cu^{++} + e^{-}$$
$$Fe^{+++} + e^{-} \rightleftharpoons Fe^{++}$$

One example of the initiation of a chain reaction is:

$$Cu^{++} + ROH \rightarrow Cu^{+} + RO^{\bullet} + H^{+}$$

Certain pigments and inks, *e.g.* copper acetate pigments and iron gall ink, are particularly notorious for their ability to produce metal-catalysed degradation of paper and damage can be so severe that the area covered by colourant falls out of the sheet and the area is lost. During industrial oxidative bleaching of paper pulp, the presence of traces of iron and copper can cause extensive damage to the cellulose and measures are taken to avoid this.

Loss of strength of the paper is caused during oxidation as chain scission is produced. During oxidation, carbonyl ($>C{=}O$) groups can be formed on some of the carbon atoms in the cellulose molecule.

5   DEACIDIFICATION

It is believed that the water-soluble acidity in paper can be washed out using water. Alternatively, if water is thought to be too hazardous to use on the paper, then the acidity may be neutralised while still in the paper, possibly by using a non-aqueous deacidifying treatment. Not all the acidity can be washed out of oxidised paper as some of the cellulose will have produced carboxylic acid groups on the cellulose chain. Many conservators at this stage of a conservation process prefer to introduce some alkaline material into the paper to neutralise any acids not washed out and which will remain behind to neutralise any acidity that may develop in the future.

Barrow's examinations of stable, old papers found that most had a significant calcium or magnesium carbonate content. Today, some of the best aqueous deacidification treatments introduce these compounds. There are several popular materials used for aqueous deacidification, one of these is calcium hydroxide solution (lime water). Calcium hydroxide is not very soluble in water; a saturated solution contains $1.85 \, g \, L^{-1}$ in ambient conditions and has a pH of 12.4 which is quite high. Severely degraded papers are suspected to undergo alkaline hydrolysis and some inks can bleed at such a high pH. While the paper is in the solution, the acids are neutralised. When treated paper is removed from the solution, the imbibed calcium hydroxide reacts with carbon dioxide in the air to form stable calcium carbonate. If the object has a very black area, *e.g.* as in some mezzotint prints, the dried precipitated carbonate may sometimes be visible as a white bloom, however, usually it is not seen. The eventual pH of the paper is in the range 8.0–9.0.

If carbon dioxide is bubbled through the calcium hydroxide solution, the carbonate is precipitated at first but later it dissolves again with the formation of calcium bicarbonate solution. This solution varies in pH depending on the quantity of carbon dioxide in solution but the pH is generally about 7. It can be used as a deacidifying treatment when a high pH must be avoided.

A process called the Barrow two-stage treatment deposits twice as much $CaCO_3$ by first immersing in calcium hydroxide, then calcium bicarbonate:

$$Ca(OH)_2 + Ca(HCO_3)_2 \rightarrow 2CaCO_3 + 2H_2O$$

Barrow's method is seldom used now, as the quantity of $CaCO_3$ deposited with the single step processes is considered adequate.

Magnesium hydrogen carbonate may be prepared by bubbling CO_2 through a suspension of magnesium carbonate (light) in water. Many conservators make this by placing the suspension in a soda siphon and screwing on a bulb of compressed CO_2. This has a neutral, if not slightly acidic, pH. The eventual product, on drying, is a form of basic magnesium carbonate.

Currently the calcium-based treatments are preferred over those containing magnesium as the former seem to induce less yellowing on ageing. Alkaline compounds of Group I have never found general favour as deacidifying treatments as they cause yellowing, are too alkaline or are hygroscopic, however, a few workers have recommended slightly alkaline sodium salts, *e.g.* borates, as buffers.

If a paper becomes extremely fragile when immersed in water or has dyes or image materials that are water-soluble or otherwise fugitive, immersion in water cannot be used for deacidification. One option is to use one from a range of non-aqueous treatments. Calcium and magnesium are both in Group II of the Periodic Table. The next heaviest element in the group is barium, which has a hydroxide, and unlike those for Ca and Mg, is soluble in methanol. $Ba(OH)_2$ solution was used for deacidification in the period 1970–1985 (approximately), but fell out of use as health and safety concerns about the toxicity of Ba compounds became more important.

Historically, the next successful agent to be used was magnesium methoxide, $Mg(OMe)_2$, which can be also be dissolved in methanol. The solution could be diluted in fluorocarbon solvents that have the advantage of being poor solvents, thus reducing the tendency for dyes and inks to bleed from the paper that might have occurred if further methanol had been used as a diluent. However, the solution proved too water-sensitive for many purposes as the $Mg(OMe)_2$ tends to react readily with water and precipitates out as $Mg(OH)_2$. This problem was overcome by saturating the solution with CO_2. The result is now a popular product called methyl magnesium carbonate, although its precise composition is uncertain. The fluorocarbon diluent has now been replaced by environmentally safer solvents.

A different approach is to use an inert organic solvent that contains microparticles of an alkaline material such as calcium hydroxide or magnesium oxide. These treatments are suitable for single sheets where the suspension can be sprayed on or the sheet immersed. Another process uses micro-particles directed at the paper in a stream of air. Until recently, it was thought that this type of treatment could not work, as without water the base and acid do not have an easy time making contact to bring about a neutralisation reaction. However, the processes perform well in accelerated ageing tests and we can only assume that paper contains sufficient water in ambient conditions for acid to move through fibres until they make contact with one of the neutralising particles. Several other ingenious methods for deacidifying paper are presently available.

Many libraries have large holdings of books printed on acid paper. Deacidifying books individually is a very time-consuming and expensive process. A process in which large numbers of books could be deacidified at the same time (mass deacidification) is a desirable invention. The problem has

produced much research and some novel deacidifying agents have been produced. One historically important and effective method was the use of diethyl
zinc. This chemical is a gas at ambient temperature and pressure, and has to
be used in large vacuum chambers. Since the gas reacts with water, the books
first have to be dried for 3 days before the deacidifying gas is allowed to diffuse into the volumes. Next, excess diethyl zinc is destroyed by the addition
of ethanol vapour. Moist carbon dioxide is finally introduced into the chamber to ensure the alkaline reserve is converted to zinc carbonate.

Cyclohexylamine carbonate, morpholine, amines in general and ammonia
itself have all been considered as deacidifying treatments. Many are liquids
with a significant vapour pressure or are gases, which makes them potentially
useful for mass processes. However, all suffer from the problem that the
deacidification produced is not permanent and some cause yellowing of the
paper. None is currently in favour.

6   BLEACHING AND WASHING OF PAPER TO REMOVE DISCOLOURATION

Discoloured paper can be made lighter by washing or bleaching. Uniform
discolouration is nowadays considered much more acceptable on paper than
it was 20 years ago and consequently, less washing and very much less bleaching is performed. Non-uniform discolouration is aesthetically displeasing and
these techniques may be used to lessen the effect. Washing in water removes
some discolouration but usually only a small fraction of that is possible by
bleaching. Higher temperatures produce faster and more complete washing.
The addition of a surfactant can also help especially with a well-sized
(hydrophobic) paper. Many different washing processes are performed by
conservators because the image materials can be fugitive in water, *e.g.* some
watercolour paints will become mobile when immersed in water. In such a
case it is desirable to carry out washing while exposing the paint to the minimum possible amount of water. When fugitive pigment is present, simple
immersion of the paper in a bath of water may be unsuitable but a process
where the paper is placed on a suction table and exposed to ultrasonically
generated water mist might be much better as the mechanism for movement
of particles through liquid water is reduced.

On ageing, paper components degrade and may produce double bonds in
the constituent molecules. When a few double bonds are present and they are
present as alternating double and single bonds, they are called conjugated
double bonds. When there are sufficient conjugated double bonds they start
to absorb visible light and they become coloured. Organic dyes and most discolouration in paper owe their colour to this phenomenon. If some or even
one of the double bonds can be destroyed, the paper may stop absorbing visible

light as the conjugation is reduced or destroyed and colour is removed, *i.e.* bleaching. In paper conservation bleaching, double bonds are usually destroyed by oxidising agents, less frequently by reducing agents. If discolouration returns several months or more after bleaching, this is known as reversion. The presence of carbonyl groups, produced during oxidative bleaching, is believed to be a cause of reversion.

In medieval times, undyed cellulosic textiles were bleached by "grassing" them, *i.e.* they were laid out on fields of grass. The combined effect of morning dew and the sun being allowed to shine on them for several months, produced bleaching. The invention of cheap ways of producing chlorine made grassing redundant after 1774. An adaptation of the slow, but effective, method of grassing has been used for the last few decades for bleaching discoloured paper. The paper is immersed in water in a shallow tray and the dish is covered with a sheet of UV-absorbing Perspex and laid in strong sunlight or under a bright artificial light. After several hours, the bleaching is complete. Many conservators use a slightly alkaline solution of magnesium hydrogen carbonate instead of pure water.

The mechanism for bleaching is not fully understood. Probably the discoloured areas absorb light and transfer the energy to the oxygenated water which then produces oxidising species, *e.g.* singlet oxygen or peroxy radicals, which then destroy double bonds and hence the colour. A problem with other bleaching treatments that are based on the use of oxidising solutions is that the clean paper fibres are oxidised as well as the discoloured ones, resulting in unwanted chain scission and oxidation. The advantage of light bleaching is that as the discoloured areas are the only ones that absorb light, they are the only areas that are oxidised.

Many of the bleaching processes are derived from the paper industry as paper pulps often need to be bleached before being formed into paper products. Much more research has been carried out on bleaching than has ever been done by conservation scientists and the paper science literature is a good place for any paper conservation scientist wanting more information.

When chlorine comes into contact with either sodium, potassium or calcium hydroxide, the corresponding hypochlorite is formed. The hypochlorite ion (OCl^-) can oxidise double bonds in discoloured paper producing bleaching. Sodium and potassium hypochlorites are available in an aqueous solution and are highly alkaline because excess hydroxide is present. Undiluted, the solutions are so alkaline that swelling of the cellulose and unwanted alkaline degradation may occur. Of the hypochlorites, it is usually safest to use calcium hypochlorite, also known as bleaching powder, as, in practice, it produces less alkaline conditions. Bleaching powder has to be dissolved in water prior to use.

The best type of bleaching occurs when there is least fibre damage but adequate bleaching. The oxidising bleaches can be ranked for their damage to paper fibres, with the main reactive species given in brackets. In order of increasing damage they are:

- alkaline stabilised hydrogen peroxide (HOO$^-$/HOOH),
- chlorite/chlorine dioxide (ClO$_2$$^-$/ClO$_2$),
- pH 9.0 hypochlorite (OCl$^-$),
- pH 4.5 hypochlorite (Cl$_2$/HOCl), and
- pH 7.0 hypochlorite (HOCl/OCl$^-$).

In hypochlorite solution, three species are present which can produce bleaching: chlorine, hypochlorite ions (OCl$^-$) and hypochlorous acid (HOCl). In pH neutral conditions, HOCl is produced. This is considered the most harmful species for bleaching as it produces aldehyde groups in oxidised cellulose that may later produce yellowing on ageing. The safest bleaching occurs above pH 9.5–10.0. When the hypochlorite ion is the dominant bleaching species, the damage to cellulose is least and the product of oxidation is the carboxyl group that does not cause colour reversion on ageing. When initially bleaching an acidic paper, care should be taken that the paper does not alter the pH of the bleaching solution outside the desired range. Burgess gives details of bleaching procedures.

Hydrogen peroxide is a versatile and popular bleach in both textile and paper conservation. Research in the paper industry has shown that although hydrogen peroxide is a good bleach, higher levels of cellulose degradation can occur where there are traces of metal contaminants, *e.g.* iron and copper. Alkaline hydrogen peroxide gradually decomposes-evolving oxygen bubbles, and this can be an unwanted and damaging side-effect with some objects. Although, many conservators use hydrogen peroxide solutions in the range 0.5–3.0% just made alkaline with a few drops of ammonia solution, others prefer to use a fully buffered and stabilised solution. Such recommended solutions for bleaching at pH 8–9 may be made quite complex containing magnesium sulfate, sodium silicate, hydrogen peroxide and a buffer solution containing sodium phosphate, boric acid or sodium carbonate for pH 8, 9 and 10 baths, respectively.

Hydrogen peroxide is mostly used in aqueous solutions but it can be dissolved in diethyl ether. If hydrogen peroxide aqueous solution is placed in contact with diethyl ether, the ether layer floats on top as it is less dense. On shaking the layers together, some of the hydrogen peroxide goes into the ether. The pigments, lead white (basic lead carbonate) and red lead may react with hydrogen sulfide in ambient air and blacken by the formation of a thin, black, lead sulfide layer. The sulfide can be oxidised by applying the ethereal solution of hydrogen peroxide when the lead sulfide forms white lead sulfate and the original appearance is usually regained. The desired reaction occurs with both aqueous and ethereal peroxide, but the ethereal solution has the advantage that no cockling, swelling of paint or local redistribution of water-soluble materials occurs.

Sodium borohydride is the most widely used of the reducing bleaches. Short treatments reduce aldehyde and carbonyl groups back to alcohol groups and cause some bleaching and stabilisation of the cellulose against future ageing. Longer treatments reduce carbon–carbon double bonds by adding hydrogen to them. Two other borohydride salts can also be used, the tetraethylammonium and tetramethylammonium borohydride, however, sodium borohydride is the most effective bleach. All three salts are soluble in water and ethanol. Unfortunately, borohydrides decompose in water with the evolution of hydrogen bubbles with associated possible physical damage to the paper and image. The most stable solution is the tetraethylammonium salt. Certain metals present as impurities in the paper or as pigments, such as iron, copper and manganese, can catalyse this decomposition. The borohydrides are used in the concentration range 0.01–2.0% and produce solutions of about pH 9.

7   FOXING

Old paper may appear uniformly yellowed due to age, but another phenomenon is foxing where yellow/brown spots appear on the paper. For many years, a controversy raged as to whether the spots were caused by microbiological agents or were tiny particles of iron in the paper that had corroded to brown iron corrosion products. The truth is that both explanations are valid. Under ultraviolet light, the two types of foxing can be distinguished. Spots that fluoresce brown, and may have a pale blue edging, are caused by fungi; some spots in this category appear pale blue in UV and are invisible in daylight. Foxing caused by iron particles appears black against the fluorescence of the paper.

Iron particles are believed to have been introduced to the paper from the water used for the papermaking process or come from the machinery used. The latter explanation has some validity as there are some papers with brass (Cu/Zn alloy) particles in them; these have often corroded to black or green copper corrosion products that may include chlorine from residues of bleaching agents.

Microbiological foxing is formed very slowly and is thus difficult to reproduce artificially in the laboratory. Besides, identifying the fungi responsible by sampling it from paper is fraught with problems as the samples always include other species in the paper. Whatever mechanism is causing foxing, the production of spots is slowed down or prevented by storage in low RH conditions.

8   ACCELERATED AGEING TESTS

When conservation treatments are being developed it is desirable to reproduce the effects of subsequent ageing to see if the treatment has stabilised or
destabilised the treated material. Similarly, conservation materials, *e.g.* adhesives, packaging, *etc.*, need to be aged to see if they will develop undesirable
characteristics on ageing, *e.g.* discolouration, brittleness, shrinkage, acidity, *etc.*
In order to find out how papers will respond to ageing, several techniques
have been developed. The method used should reflect the conditions to which
the paper will be subjected. If a paper is subjected to a great deal of light,
light ageing should be used, however, if the paper is stored in the dark or low
lighting conditions, a heat ageing regime is more appropriate.

The vast majority of paper in libraries, archives and museums is kept in
conditions of total darkness or in low incident light that contains little or no
ultraviolet light. In such conditions, light degradation is of no or little importance. However, objects on display or which form part of a historical room
may receive sufficient light to cause serious concerns about degradation. A
few dyes and pigments are extremely sensitive to light and may need to have
a ban on being exhibited at all.

In this section, we will concentrate on heat ageing which is intended to
reproduce the effects of ageing in the dark and in the absence of air pollutants. Chemical reactions in general, and those which cause deterioration of
paper in particular, slow down when the temperature is decreased and accelerate when the temperature is increased. In response to this fact, some libraries
and archives have built stores that can maintain a low temperature to prolong
the life of their collections. There is an equation that was defined by Arrhenius,
which can be used to predict the effects of temperature on the rate of chemical reactions/ageing, here it is:

$$k = A\,e^{-E_a/RT}$$

where k is the rate of reaction, A and R are constants, E_a is an experimentally
derived value called the activation energy and T is the absolute temperature
(add 273 to the temperature in degrees Celsius to get this). Only a few paper
scientists need to perform calculations using this equation, but its importance
to most conservators and paper conservation scientists is that it describes the
way that the rate of reactions accelerate by two or three times for every
increase in temperature by 10°C.

Barrow (see above) used ageing in an oven for 72 hours at 100°C to replicate
the effects of natural ageing of paper for 25 years. Nowadays, this assumption
is still used, but regarded as simplistic and not applicable to all types of paper.
Many laboratories prefer to perform their paper ageing in an oven at a controlled
relative humidity and at a lower temperature. There is no consensus, but a

typical set of conditions is 80°C and 65% RH for 4 weeks. Usually, there is no attempt to offer an equivalent time for ambient ageing. Recent research from several independent laboratories has shown that the degradation products from paper ageing appear to accelerate the ageing. Accordingly, some scientists recommend ageing paper in sealed glass tubes which retain degradation products and keep the water content of the paper approximately constant. As a general rule, the closer the conditions of accelerated ageing are to ambient storage conditions, the better the results will be. The only drawback in ageing at lower temperatures is that the time needed for accelerated ageing is increased.

Accelerated ageing using light should only be used to anticipate the effects of light on paper and not for the effects of ageing generally. The energy of photons in light transfers energy to paper in a different way and intensity to the thermal energy of heat ageing and the results of one type of testing may be entirely different from the other. A paper that yellows in thermal ageing may bleach in light ageing. White light contains light of different colours; the red end of the spectrum contains light of low energy, and the packets of light (the photons) are not capable of damaging paper. At the blue end of the spectrum, the photons contain more energy. Some illuminants contain ultraviolet light which is even more damaging to the paper. Accelerated ageing should be performed with light with a spectral distribution as close as to that in the intended storage conditions.

Fluctuating relative humidity can be bad for paper because of the dimensional stability of the paper, *i.e.* production of cockling. However, recent work has shown that large losses of tensile strength can result from a regime of wide changes in relative humidity. This promises to be an interesting field of research in the future. Elevated temperature combined with fluctuating RH has been used in ageing iron gall ink on paper.

9   SAFE ENVIRONMENTS FOR PAPER

The safest conditions for storage of paper may be easy to define, but are difficult to achieve in practice. Oxidation may be eliminated by storage in nitrogen but this is not practical in all but a few instances. Recommended conditions are a compromise between what is ideal and realistic.

At RH greater than about 65%, fungal growth may occur, so the RH should be below this. Air circulation is an additional advantage. At low RH, deterioration reactions that need water to proceed will be slowed down but paper and glue become more brittle, so 40% is the effective lower limit for general storage. Chemical reactions slow down with decreasing temperature, so ideally, storage should be at as low a temperature as possible, but scholars do not like to work in cold conditions and, anyway, the depressed temperatures are expensive to maintain. International standards give the temperature for storage as about

18–20°C. On exhibition, light levels are recommended to be as low as possible (50–150 lux usually) with no UV or at least less than $75\,\mu\,$Watts lumen^{-1}. Some paper-based objects may need a change in these recommendations, *e.g.* colour photographs are generally stored at lower temperatures.

Materials for the repair of paper should be tested by accelerated ageing to ensure that they will be same after they have been applied to objects. There are several desirable features for conservation materials; they should be easily removable again; they should not discolour; lose flexibility; be or become acidic or emit a substance harmful to artefacts in the collection. A few examples will illustrate the hazards of using bad materials. Inappropriate storage materials can contribute to the deterioration of paper. Bad quality mounting boards can contain acidic or coloured materials that can migrate to works of art and discolour them. Some adhesives can become so insoluble in common solvents that they are impossible to remove. In the case of rubber-based adhesive tape, the adhesive becomes totally insoluble and brown. Some poly(vinyl acetate) adhesives emit acetic acid when they age and this can act as a catalyst for the corrosion of lead in seals attached to documents. Some papers emit aldehydes and peroxide that cause fading of black and white photographic images by attacking the silver micro-particles.

10   METHODS FOR MONITORING THE DETERIORATION OF PAPER

10.1   Physical Measurements

The deterioration of paper is manifested in several ways, for example, extensive discolouration and loss of mechanical strength can occur. The early detection of changes and their measurement needs special apparatus.

Colour change on dark ageing produces discolouration of the paper to yellow/brown. Light ageing similarly produces this type of discolouration but may, less commonly, produce bleaching. Not only may dyes and pigments in the paper bleach, but initially discoloured paper may also become whiter. Light spectroscopy can be used to measure the entire visible reflectance spectrum of the paper, but usually the reflectance at a single wavelength near the blue/violet end of the spectrum is used; 400 and 457 nm can be found in the conservation literature. Possibly the most convenient method is to use a tristimulus colour measuring meter which reports the hue as three parameters; many systems for colour measurement have been devised, at present the most used is the CIELAB system. Total colour change or a change in b^* (which represents yellowness) have been used when studying discolouration.

Tensile strength is one of the commonly measured mechanical properties. Tensile testing machines are very versatile and used for all types of conservation research work, as they can be used to pull apart and thereby measure the strength of yarns, adhesive joints, plastics, *etc*. They can also be adapted to crush materials such as stone. In tensile testing of paper, a strip is clamped between two jaws and is pulled apart as the jaws start to separate. A load cell measures the stress on the paper. As the paper stretches, the stress increases until the paper fibres start to slip over one another or the paper starts to tear, in which case the stress will decrease. The graph of stress against elongation can be used to calculate several useful parameters relating to the strength and elasticity of the paper.

When paper breaks under tension, the break involves a combination of fibres breaking and bonds between fibres being disrupted. In very strong fibres, the paper breaks mostly by failure of the fibre-to-fibre bonds, while in weak papers the break occurs by failure of the fibres themselves. If the jaws holding the paper strip are very close together, the fibre strength dominates the results obtained as the contribution from fibre-to-fibre bonding is minimised. This is called zero span tensile testing.

In all the mechanical tests described in this section, relative humidity must be controlled carefully as the properties measured change significantly. International and national standards will give the required RH history for samples to be tested. As there is a hysteresis effect with water content and RH of paper, it may be important that the desired RH for testing is approached from one of either the dry side or the damp side. Of course, different results will be obtained if properties differ between the CD and MD, but also some properties, *e.g.* gloss, colour, roughness, *etc*., may differ from one side of the sheet to the other. However, a major problem with all the mechanical tests is variability of results. Not only is there non-uniformity in the paper, but test results which involve breakage of the sample have a scatter in values which is larger than is desirable because of the catastrophic failure mechanism. Because of the scatter, a large number of samples have to be tested to obtain a result that is statistically useful. Testing 15–30 strips is usual. This consumes a large amount of sample and this type of test is not so useful for irregular papers or those in short supply such as those from objects.

The ideal mechanical test would be one in which the properties related to the use of the paper were tested. Although weak papers have low tensile strength it may be of more interest to measure how they behave when folded. A test that measures the tensile strength after one fold has been proposed.

If folding is of paramount importance, the MIT folding endurance test is of great value. In this test, the test strip is held vertically under tension. The upper jaw that holds the paper is fixed but the lower jaw swings like a pendulum, with the centre of rotation being the point where the paper enters the jaws. The tips of the jaws are curved in a defined manner. A counter measures the number of folds required before the paper breaks. A strong paper will endure over a hundred folds but very weak papers will only endure a few folds and for these the test is not suitable.

10.2 Chemical Methods

The alternative to measuring the physical attributes of paper is to measure the chemical properties of the paper. As with physical tests, care has to be taken that the sample of paper is representative of the entire sample, for example, in an old book the edges of a book block are often more degraded than the centre because the edge has absorbed air pollutants and is often of higher acidity. The popular methods will now be briefly described, readers should see fuller descriptions before attempting any of these methods.

Although some papers are acidic when new, ageing generally increases acidity and pH can be used to monitor the progress of ageing. There are several methods for measuring the pH of paper including pH-sensitive dyes that can be placed onto the paper, and surface pH electrodes that can be placed directly on the surface of the dampened paper. However, the best method is to macerate paper in pure water and measure the pH of the aqueous extract after a period. There are several standard methods that are similar. Typically, they mix 1 g of paper with 50 mL water (hot or at room temperature) and measure pH after 1 hour.

The solubility of paper carbohydrates in sodium hydroxide (NaOH) of various strengths is a useful method for both characterising new paper and following degradation. One method uses a 17.5% solution of NaOH to treat paper, so that, of the carbohydrate component, only cellulose of DP > 200 remains undissolved, and lignin and mineral fillers remain largely unaffected. The insoluble carbohydrate is called α-cellulose. The α-cellulose content will decrease on ageing. After the 17.5% NaOH treatment the solution contains β- and γ-cellulose. The β-cellulose, which has a DP of 14–200, is soluble until the alkali is neutralised, while the γ-cellulose remains in solution after neutralisation; it has a DP of less than 10. Materials other than paper carbohydrates, *e.g.* lignin, interfere with the results.

A related method involves measuring the solubility of the paper in 1% NaOH at 100°C after a period of time, *e.g.* 60 min. The solubility is proportional to the copper number (see below) that is itself proportional to the number of carbonyl groups in the paper, a measure of the oxidation and hydrolysis of the paper. This method has also been used on lignin-containing fibres of various kinds with some success.

The copper number of paper may be obtained using copper(II) solution, some of which is reduced to Cu(I) oxide as the carbonyl groups are oxidised to carboxylic acid. There are several variants of the test, one of which involves treating the produced Cu(I) oxide with phosphomolybdic acid. The reduced molybdenum is then titrated with potassium permanganate.

Phenylhydrazine is able to react with carbonyl groups to give coloured compounds. Paper sheets can be reacted with this reagent or modifications of it (*e.g.* 4-nitrophenylhydrazine) and the colour produced can be measured by reflectance spectroscopy. The colour may also be measured by dissolving the paper and measuring the absorbance of the solution. Lignin interferes badly with this test.

A test that measures the quantity of carboxyl groups uses methylene blue absorption. A sample of the paper is immersed in a solution of methylene blue and the absorbance of the dye solution at 620 nm is measured before immersion of the paper and at the end of the experiment. The amount of sample needed to produce 50% exhaustion of the dye solution is used to calculate the carboxyl content.

Chemical reagents can dissolve cellulose. Three of these are aqueous solutions containing copper ethylenediamine, cuprammonium hydroxide or cadmium ethylenediamine (cadoxen). The solutions all cause some damage to the cellulose by alkaline degradation during dissolution. Some workers perform a pre-treatment in which oxidised groups are reduced, which decreases the amount of alkaline degradation considerably. Measurements of cellulose solutions in the copper-containing reagents should ideally be performed in the absence of air as oxygen can cause degradation of the cellulose. The cadmium-containing reagent is very toxic and for that reason is less frequently used than it used to be, however, any reader of the paper conservation research literature will find it mentioned more often than the others mentioned here. The higher the DP of the cellulose, the greater is the viscosity of the solution.

Although DP can be determined by measuring the viscosity of cellulose solutions, this is a skilled business and needs much practice. For those laboratories with the money for expensive equipment, a better option may be size exclusion chromatography (SEC). Much work in conservation research has been carried out in cadoxen solutions, however, cadoxen is aggressive to SEC columns and more recent work has used LiCl/dimethyl acetamide. The size exclusion column holds back larger molecules and when the cellulose emerges from the column, the smallest molecules come out first. A detector that measures the amount of material emerging from the column enables the user to see not only the average DP but how the DPs are distributed about the average. The column can be calibrated with solutions containing carbohydrates or other polymers of known DP.

Fourier transform infrared (FTIR) spectroscopy produces a spectrum for paper with many peaks. Each peak corresponds to a vibration in a chemical bond. The frequency of infrared peaks is usually measured in the unit cm^{-1}, the number of wavelengths which can be fitted into a centimetre. Peaks can be attributed to types of chemical bond; for example, a peak at $2900\,cm^{-1}$

corresponds to the stretching of C—H bonds, while a peak at $1740\,cm^{-1}$ corresponds to a stretching of a carbonyl group. As the strength of the C—H peak remains more or less constant during the ageing of paper, the ratio of the strength of the carbonyl peak with the C—H peak can be used to follow the oxidation of the paper. The ratio of the peaks at $1372\,cm^{-1}$ (C—H) with that at $2900\,cm^{-1}$ can be used to monitor the crystallinity of the cellulose.

The FTIR spectrum of a piece of paper can be measured non-destructively using an FTIR-microscope or, probably more reproducibly, by taking a sample and either placing it directly in a diamond cell or grinding it with infrared-transparent potassium bromide and pressing it into a disc to go into the spectrometer.

11 CHARACTERISATION OF PAPER

There are many parameters that can be used to describe paper, and this section describes some of the scientific ones often used. Paper is made of cellulose fibres. Some fibres have characteristic features when viewed on a microscope and an experienced person can identify the source of fibres in a paper sample and maybe even estimate the percentages in a mixture. Identification is greatly facilitated by macerating the paper and separating the fibres. There are many techniques for doing this, but many involve adding a few drops of alkali or boiling the paper in an aqueous solution. Although some scientists prefer to use a scanning electron microscope, most fibre identifications are performed on a polarising light microscope as interior features can be seen. Fibres are laid out on a glass microscope slide and mounted under a cover-slip using a liquid mountant. Several features of the fibre, and the associated botanical debris, are used by the microscopist who may also use staining techniques. Many fibres can be easily identified, *e.g.* cotton and linen, and it is easy to tell if the fibre is coniferous wood, however, it is more difficult to distinguish different wood-derived fibres from one another. The traditional Oriental paper fibres, *e.g.* kozo, mitsumata, gampi, *etc.*, are very difficult to differentiate.

Many standards describing paper to be used for conservation purposes will specify that the papers should be free of alum and lignin. Several spot tests can be used to detect these and other materials found in paper; one of each will be described. One test for alum uses a solution of a chemical called aluminon. The pale pink solution is applied to the surface of the paper and allowed to dry. A pale pink spot is produced when no aluminium salts are present; a darker pink or red is seen when such salts are present. Lignin can be detected using a freshly prepared solution of phloroglucinol in hydrochloric acid. A drop of the colourless solution is applied to the paper and a red colouration indicates the presence of lignin. There are several tests for rosin.

In one of these, the Raspail test, a drop of concentrated sugar solution is added to the paper, blotted off after 1–2 min and a drop of concentrated sulfuric acid is added. Rosin gives a red colour.

Two types of traditional sizes, namely starch and glue (protein), can be tested for, and the tests are also used for identification of adhesives on paper artefacts. Starch can be detected by placing a drop of iodine solution onto the area to be tested, and a blue–black colouration reveals the presence of starch. There are many tests for protein, but in the ninhydrin test a solution of ninhydrin in methyl Cellosolve with a trace of surfactant is added to the paper that is then heated in an oven at 100°C. A purplish stain indicates the presence of protein.

12   CONCLUSIONS

Paper is a complicated material. Besides lacking homogeneity on a macro- and micro-scopic level, it contains many other materials such as image media or paper constituents. It is not practical for the paper conservator to examine all the components and properties of a paper object before it is treated. However, experience of treating a wide variety of paper objects and knowledge of the properties of paper and associated materials enables the conservator to achieve success. This same basic scientific knowledge enables the conservator to communicate effectively with other conservation scientists who can investigate problems outside the conservator's area of specialisation. It is hoped that this text has gone some way to providing this information.

REFERENCES AND FURTHER READING

B.L. Browning, *Analysis of Paper*, Marcel Dekker Inc., New York and Basel, 1997.

BSI: Recommendations for the storage and exhibition of archival documents, BS5454: 2000, British Standards Institute, London.

BSI: Repair and allied processes for the conservation of documents – recommendations, BS4971: 2002, British Standards Institute, London.

H. Burgess, Practical considerations for conservation bleaching, *J. Int. Inst. Conserv. – Canad. Group*, 1988, **13**, 11–26.

G. Petherbridge, *Conservation of Library and Archive Materials and the Graphic Arts*, Butterworths, London, 1987.

J.C. Roberts, *Paper Chemistry*, Blackie, Glasgow and London, 1991.

M. Strlič and J. Kolar, Ageing and Stabilisation of Paper, National and University Library, Ljubliana, 2005.

The journal Restaurator is an excellent starting point for reading about the latest developments in paper conservation science. The Paper Conservator is similarly useful but with a greater emphasis on practical treatments and case studies.

CHAPTER 4

Textiles

PAUL GARSIDE[1] AND PAUL WYETH[2]

[1] AHRC Research Centre for Textile Conservation and Textile Studies, and Textile Conservation Centre, Winchester School of Art, University of Southampton Winchester Campus, Park Avenue, Winchester SO23 8DL, UK
[2] Textile Conservation Centre, Winchester School of Art, University of Southampton Winchester Campus, Park Avenue, Winchester SO23 8DL, UK

1 THE VARIETY OF TEXTILES

According to the dictionary definition, a textile is 'a woven fabric or any kind of cloth', leading us to think of fashionable costume, for example ceremonial banners and interior furnishings, however, this belies the range of amazing artefacts and further fascinating challenges that are routinely presented to a textile conservator.

To illustrate the point, the Textiles section of Conservation News (May 2004), the magazine of the United Kingdom Institute for Conservation of Historic and Artistic Works, carried reports on three international meetings held in 2003. The reviews highlighted something of the wonderful variety of textile artefacts: Henry VIII's tapestries, the Bishop of Winchester's Gothic ankle boots, 17th century doublets, the original American Star Spangled banner, English Trades Union banners and 18th century Parisian furniture upholstery.

Textiles then represent a vital element of our material legacy, each with a tale telling of international trade, social history, agricultural development, artistic trends, technological progress and so on. Yet such stories are easily lost. Textiles are ephemeral artefacts. They are composed, in the main, of organic materials which decompose naturally under the influence of the environment, degradation leading to the weakening of fabrics, embrittlement of the fibres and fragility.

It is the job of the textile conservator to ensure, as far as is possible, the longevity of such objects, for all to enjoy. The conservation scientist can help to inform the decisions that the conservator must make about the treatment of the objects. The constituent materials have to be identified, requiring both traditional approaches and the application of modern analytical technology, and an appreciation given of their current physical and chemical state.

In this chapter, we will describe the make-up of just three textile materials, all natural fibres, and will further present the mechanisms of degradation of their principal components. Our selection of two of them is somewhat indulgent as they are each the focus of current research projects. However, since all three compose the fabric of key historic textiles which were recently the subjects of commissions completed by Conservation Services at the Textile Conservation Centre, our choice was all the more easily made. Here we are keen to place the science in context, and so use these artefacts to introduce the fibres and their chemistry of ageing through a conservation science perspective.

2 TEXTILE MATERIALS

Before moving on to the case studies though, something more about the nature of textile materials and a potted history of textile fibres. As we have already suggested, there is often much more to an object than just the textile component which in itself can be complex. Presented with the banner of the Amalgamated Society of Locomotive Engineers and Firemen, Nine Elms Branch, made by the Victorian firm of George Tutill and Company, conservators had to consider not only the silk fabric but also the painted images on front and back and the pole from which it hung. They also needed to know about the nature of the disfiguring bloom on the surface of the paint and how this could be best removed. They had to have knowledge of the possible cleaning protocol, which could have involved the use of buffered surfactant solutions, lytic enzymes or the application of organic solvents. In consolidating damaged areas, they also required advice on the optimum choice of adhesive for attaching support and fine netting. Science then had a crucial rôle to play.

Since this book is devoted to the description of the materials from which objects are made, we can simply focus on the textile components. Even then, for a single textile, this can still afford a wealth of different types, as evidenced by the third case study, the *Tree of Jesse* a rare 15th century Rhenish tapestry, woven in wool on linen support yarn with stunning details worked in colourfully dyed silk and golden threads. There are over a dozen types of metal thread including gold foil, and gold powder on paper or animal gut strips wrapped around silken cores. Before getting carried away with the incredible craftsmanship of gold wrapped threads, which dates from Roman times, or the brilliant hues of mordanted madder or other natural dyes, we must readjust our focus and return to the organic fibres themselves.

Seed fibres from cotton and stem fibres from flax were probably the first to be commonly woven into textiles. Linen (from flax) was the every-day fabric of ancient Egypt, and the plant was the first cultivated source of textile fibres in Europe. While cotton only became popular in Europe a few centuries ago, its history in Asia and the Central Americas dates back over 7000 years. The spinning of animal fibres into threads and yarn seems to have developed relatively late on. In Mesopotamia, woven woollen cloth was being produced just 6000 years ago, while the use of silk, in China, may only date from early in the 3rd millennium BC.

It was the search for a substitute for the luxurious but expensive silk fibres that precipitated the revolution in synthetic polymer chemistry of the last century and settled the fortunes of companies such as Courtaulds, DuPont and ICI. Nylon, polyester and acrylic threads, together with polyurethane elastomers, now constitute a large part of the textile market along with the regenerated cellulose fibres like viscose rayon and their modified counterparts, the acetates. These modern materials and even newer advanced smart and techno fibres are already finding their place in museum collections, presenting fresh challenges to conservators and conservation scientists. The topic of modern materials ageing is ripe for research.

We will leave a discourse on aspects of saving 20th century textiles until sometime in the future, when more should be known about their decay. For now we will concentrate on the fibres that make up the mainstay of the objects that currently come under the textile conservator's eye, *i.e.* the natural fibres. We can categorise these as plant fibres or animal fibres, among which there are numerous examples. We will, however, just select linen, and silk and wool respectively, choosing a single reasonably representative plant fibre, but two animal fibres since these are quite distinct. Our three exemplars will serve well to illustrate the general intricacy of natural fibre design, our depth of knowledge of the chemical composition and ordered nature of these biocomposites, and our level of understanding of the mechanisms of natural polymer ageing. Linking to the case studies will then show how such detailed awareness of the underlying science can lead to better-informed choices for the long-term preservation of our textile heritage.

3   CASE STUDY 1: LINEN FIBRES AND THE *VICTORY* SAIL

Recently, we had the privilege to work on an object of major national maritime importance, the fore topsail of *HMS Victory*. The sail from Admiral Lord Nelson's flagship was extensively damaged during the Battle of Trafalgar, fought on 21st October 1805. It was holed by musket and cannon fire, and a long gash was torn through the fabric as a mast smashed to the deck. The sail's origin can be traced back to Baxter Brothers Linen and Jute Manufacturers of Dundee; it was in use by 1803 and was probably employed at sea for roughly 18 months prior to the battle. Not only does it represent a link to one of the most crucial sea battles in British history, it is also believed to be the only surviving sail from the period. So, besides being one of the largest and most significant naval textile artefacts, it bears key witness to the early 19th century sailmakers' art.

The sail is 80 ft (24 m) wide at the foot, 55 ft (17 m) in height and weighs somewhere in the region of half a tonne. It is constructed from bolts of linen cloth (each roughly two feet wide), running from head to foot, and is reinforced with similar material running across the width of the sail in bands (Figure 1); the edge is strengthened with hemp rope. The sailcloth is of a plain weave, *i.e.* simply woven with warp and weft yarns interlacing alternately (Figure 2). In this instance the warp yarns are paired (10.8 ± 0.4 pairs cm^{-1}), running along the length of the cloth and so aligned top to bottom for the sailcloth and left to right for the bands. The weft yarns (7.9 ± 0.9 yarns cm^{-1}) run across the cloth.

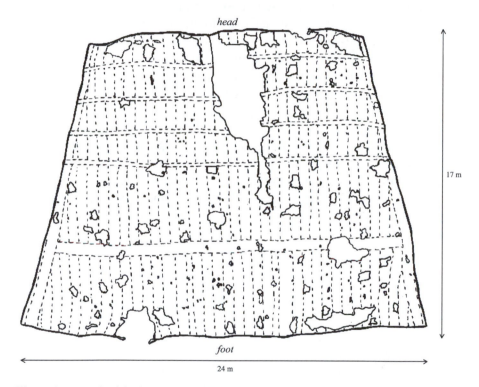

Figure 1 *A sketch of the fore topsail of HMS Victory, showing the areas of loss*

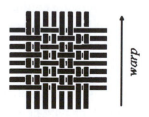

Figure 2 *Weave structure of the sailcloth, with paired warp yarns*

In 2003, the conservators at the Textile Conservation Centre were charged with carrying out first stage conservation, which involved cleaning the sail. At the same time we were asked to assess the physical state of the fore topsail, in relation to its handling, and to determine if the sailcloth was at immediate risk. To be able to provide complete answers and ultimately advise on the most appropriate approaches to conservation and display, it was essential to have an understanding of the microstructure, chemistry and degradation mechanisms of the constituent linen fibres. Further knowledge of the sail's life also proved critical – the mechanical stresses during its brief use and its exposure to the marine environment, the damage it suffered during the battle, and the more subtle deterioration over the subsequent two centuries of storage and occasional display have all played a rôle in determining its current condition.

In the following paragraphs we will describe the chemical composition of linen fibres and the microstructural hierarchy within them. We present the results of our mechanical measurements, which helped establish the current weakened condition of the cloth, and relate the reduced performance to the disruption of the fibres consequent upon ageing. This, in turn, leads into a discussion on the mechanisms of cellulose degradation (which apply equally to wood and cotton cellulose). As we will see, one expected outcome of oxidative deterioration is the generation of carboxylic acids, with potentially deleterious consequences for the cellulose polymer chains. Measuring the acidity of historic linen fibres is therefore crucial and, as outlined at the end of this case study, allowed us to make some recommendations concerning the longer-term preservation of the sail.

3.1   The Chemical Composition and Microstructure of Linen

Linen is derived from certain varieties of flax (*Linum usitatissimum*); other varieties of the species are cultivated for linseed for the production of oil. The flax fibres make up 15% by volume of the plant stem. They are found in the bast, a fibrous layer next to the woody interior, acting as structural braces. To isolate the fibres and remove the non-fibrous and woody tissue, traditionally linen straw is retted (wetted to permit biological attack on the inter-fibre glue), scutched (mechanically stripped) and hackled (combed), leaving yellow/brown ligno-cellulosic fibres. These can be bleached white for fine textiles, generally after spinning and weaving. The fibres are cellular in nature, being composed predominantly of cellulosic cell walls around a hollow lumen. Bundles of these individual cells (known as ultimates) are then cemented together by the middle lamella. The cell walls are largely made from polysaccharides (>80%), in addition to a smaller proportion of lignin (2%), proteins, pigments, waxes and minerals. The polysaccharide component is primarily cellulose (in the form α-cellulose or cellulose I) (64%), along with hemicelluloses (17%) and pectins

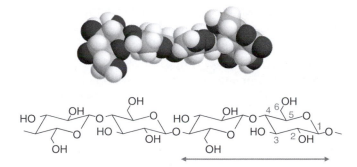

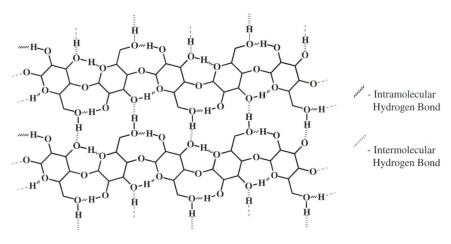

Figure 3 *Space-filling and structure diagrams of a short section of a cellulose chain. The repeat unit, cellobiose (1,4-β-D-glucopyranosyl-D-glucose) is indicated*

Figure 4a *Hydrogen bonding between cellulose chains*

(2%). Cellulose itself is a polymer built from glucose, although the repeat unit is generally taken as the dimer, cellobiose (1,4-β-D-glucopyranosyl-D-glucose; see Figure 3).

For undegraded cellulose in linen, the degree of polymerisation is typically of the order of 10,000–20,000. The straight chain nature of the polymer allows ready interaction between adjacent chains, and strong intermolecular attraction is possible, resulting from hydrogen bonding involving the hydroxyl groups (Figure 4a). This intermolecular association leads to a high degree of crystallinity (typically 70% for linen), which results in the suitability of cellulose as a structural material in plants, and also conveys a high degree of chemical resistivity (Figure 4b).

The fibres possess a complex hierarchical microstructure, in which regions of well-ordered crystalline cellulose are interspersed with areas of random

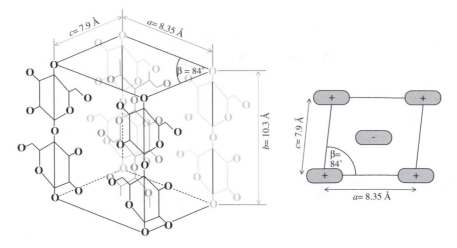

Figure 4b *The unit cell of crystalline cellulose I*

crystallinity, lacking long-range ordering, and amorphous regions, in which the ordering of the polymer breaks down completely; among these, additional matrix components such as hemicelluloses, pectins and lignin are also found.

At the molecular level, the cellulose chains are organised into parallel – and hence crystalline – bundles of 100 or so chains. These then group to form micelles or elementary fibrils, typically 5–6 nm in diameter. Aggregates of 15 or so micelles, within the amorphous intermicellar space, form microfibrils, which are roughly 75–90 nm in diameter. Bundles of microfibrils, again embedded within amorphous material, the interfibrillar space, form fibrils or macrofibrils, typically 0.5 μm in diameter. Macrofibrils are found as lamellae in the cell walls. The intermicellar and interfibrillar matrices, as well as the intercellular middle lamella, are composed of the non-crystalline components mentioned above – amorphous cellulose, hemicellulose, pectin and lignin; within these regions there are extensive networks of capillaries (Figure 5).

The cell walls are organised in layers, of which the secondary wall is the thickest (several microns), possessing the greatest degree of crystallinity and making the dominant contribution to the physical properties of the fibre. Within these walls the macrofibrils are wound around the fibre axis, with a sense and angle characteristic of the fibre species – for flax, the secondary cell wall is S (*i.e.* anticlockwise) wound at about 6.5°.

Hemicelluloses (polysaccharides composed primarily of xylose and mannose) are much shorter in length than cellulose and are extensively branched; these species may act to cross-link cellulose and other polymers, and also form the amorphous matrix of the interfibrillar space, in which cellulose microfibrils are embedded. Pectins are jelly-like, acidic polymers composed

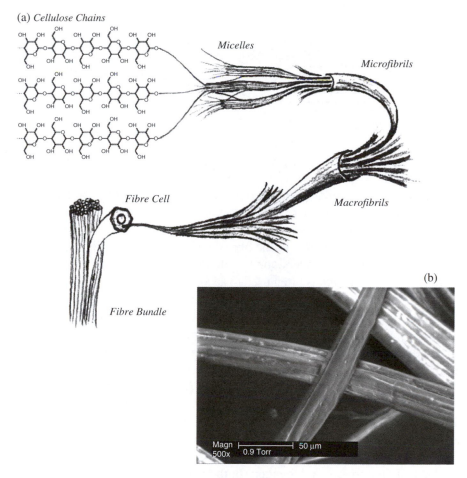

(a) *Cellulose Chains*

Micelles

Microfibrils

Fibre Cell

Macrofibrils

(b)

Fibre Bundle

Magn 500x   0.9 Torr   50 µm

Figure 5 *Hierarchical microstructure of a linen fibre: (a) from cellulose chains to the fibre and (b) a scanning electron micrograph showing the ultimate cells bundled in fibres*

of 1,4-β-(D-galacturonic acid) units, along with some galactose and arabinose; the acid groups are generally present as either methyl esters or metal salts (primarily magnesium, calcium and iron). Lignin is a small heterogeneous acidic phenylpropanoid polymer, with a complex amorphous structure; it is found within the hemicellulose matrix, to which it may be covalently bound, and is also a major component of the intercellular matrix (middle lamella). Lignin is rather hydrophobic and is highly light sensitive, particularly to ultraviolet radiation, and discolours to a deep yellow-brown hue after prolonged exposure; an increase in acidity also results.

The remainder of the fibre is composed of: waxes, which are generally found on the fibre surface; pigments, primarily chlorophyll, xanthophyll and carotene, along with other coloured material; and residual protein.

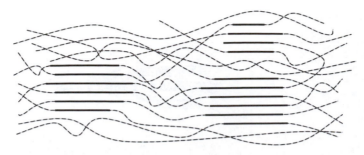

Figure 6 *The fringed micelle model of polymer microstructure*

The overall performance of the fibres is dependent on the complete hierarchical structure, from the cellulose polymer chains to the ultimate bundles. Nonetheless, it is the fibrous cellulose set within the hemicellulose matrix that is the key structural component of the fibre composite. The functional properties of cellulose can be best understood by reference to a simple model, that of the 'fringed micelle', which is often used to account for the mechanical behaviour of thermoplastics. This model presents a picture of a semi-crystalline polymer aggregate, in which nano-crystallites are embedded within an amorphous matrix (Figure 6). At the molecular level, each polymer chain passes through several ordered regions, which result from the close packing of neighbouring chains, and through several amorphous regions, in which it adopts an unordered, random coil arrangement. The strength of the material is dictated by the interlinked, inextensible crystalline component, while any visco-elastic and plastic behaviour is determined by the amorphous regions. Linen fibres exhibit elastic recovery if subjected to only a few percent elongation, but with increased loading the chains in the amorphous zones will slip past each other effecting first recoverable and then irrecoverable creep. In the extreme, of course, the aggregate will disrupt and the fibre will fracture.

3.2   The Mechanical Performance of the *Victory* Sail

Linen is particularly suited to the production of sailcloth due to its good mechanical properties, including a high strength and low extensibility; furthermore, the strength of the fibres actually increases on wetting. Inevitably, however, linen will gradually deteriorate through natural causes and its performance will reduce.

The linen fibres of the *Victory* sailcloth and, in particular, the cellulose chains are bound to have aged in the intervening 200 years. This could have progressed to the point at which the sail cannot even support its own weight. An appreciation of the sail's current condition is therefore crucial in deciding how it should be handled and may even influence the choice of display. With this in mind, we set about the mechanical testing of the cloth. However, since

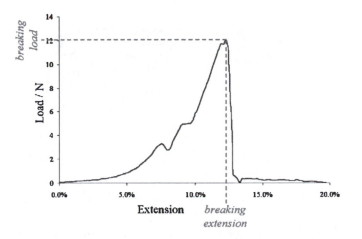

Figure 7 *A typical load/extension (stress versus strain) curve for a yarn from the Victory sail. Typically, a test length of 2.5 cm was set*

such tests are normally destructive and require free specimens, we simply had to take yarn samples from around holes in the cloth, where loose yarns could easily be removed without disfiguring the sail.

The following characteristics were measured: *linear density*, the mass per unit length of the yarns (g m^{-1}); *breaking strength*, the force required to break the sample (*N*); *elongation*, the percentage of original length by which the sample is stretched (%); and *tenacity*, the breaking strength per unit linear density (cN tex^{-1}). (By calculating the tenacities of the yarns, the problem of comparing yarns of different linear densities is avoided.) A typical mechanical test curve for a yarn, illustrating the percentage extension against load, is shown in Figure 7.

From these data, average tenacities of the yarns were calculated for the various sampling locations, as shown in Figure 8. The values are all quite close and, in any case, do not vary in a systematic manner across the sail. Some areas are more heavily degraded and weaker, but the bulk appears to have retained about 30% of its strength when new.

While tests of this sort reveal the mechanical properties of individual yarns, they do not necessarily reflect the characteristics of the cloth as a whole – the structure of the weave may also play a rôle in the load-bearing of the fabric. Since we obviously could not justify removing whole pieces of the *Victory* sail to investigate such behaviour, we employed an appropriate surrogate material. This was modern linen sailcloth of a comparable weave. Fortunately, we found a good correlation between the breaking load of individual yarns and that of the new sailcloth pieces. The only proviso was that the value for the cloth should be taken as 80% that of the fibre. The reason for this is that there is an additional failure mechanism when dealing with long specimens. For short

head

24 m

17 m

foot

Figure 8 *Average yarn tenacities (cN tex^{-1}) at various point across the sail*

samples fibre fracture and slippage of the ultimate cells (which are around 7 cm in length) are important. Slippage of the fibres themselves needs to be borne in mind as well, once the length of the yarn sample approaches that of the staple, *i.e.* the length of the fibres, which is generally a few tens of centimetres.

We were now in a position to calculate the load-bearing capacity of the *Victory* sail, and, taking account of the holes and losses at the head of the sail, could estimate, conservatively, that the sail would still be able to take its own weight and as much again, despite its aged condition. However, besides the danger of ripping the yarns and fabric, we were also concerned with the permanent stretching that could result from free-hanging display. In addition to the rôle of slippage in catastrophic failure, it can also affect such deformation; this occurs when individual fibres in a yarn slip past each other, when the ultimates within a single fibre slide over each other, and when the secondary bonds that bind polymer chains together break, allowing the chains to creep past each other before new bonds are formed. From further direct investigations we determined that the sail would undergo a permanent deformation long before it fails. We were then able to offer advice concerning the handling of the sail during its conservation to avoid undue and prolonged stress, and suggestions for its display, resting at a shallow angle on a support platform. These recommendations were necessary in light of the reduced strength and, in particular, facile deformation consequent upon the aged condition of the sailcloth and the degraded state of the constituent fibres.

3.3 Degradation of Cellulosic Fibres

Deterioration of each of the fibres' various components will affect the mechanical performance, but to different extents and through different mechanisms. Breakdown of the intercellular glue will obviously facilitate slippage of the ultimates, and this is often a particular problem for archaeological linen. Here though, we have chosen to focus mainly on the deterioration of the structural cellulose filler, which may be the more pertinent to the weakened condition of the *Victory* sailcloth.

Cellulosic materials are susceptible to a range of degradation processes, including those associated with chemical (*e.g.* acid attack, photolysis, microbial and fungal metabolism) and thermal degradation. In general, the deterioration of the material is due to oxidative processes or hydrolysis, leading to both scission and cross-linking of the cellulose polymer. In any event, the general result is a polymer assembly with a lowered degree of polymerisation, increased crystallinity due to aggregation of some of the smaller chains, and the development of some network character, and hence weakening and embrittlement.

Humidity and Heat. Moisture is crucial to the normal behaviour of cellulosic fibres. Under moderate conditions (relative humidity 45–65%) water is readily absorbed through the network of pores running through a fibre cell, it coats cellulose crystallites and acts as a plasticiser of the amorphous regions, disrupting inter-chain hydrogen bonds. Without this bound water the fibre would be permanently brittle, with an effective glass transition point way above room temperature.

Humidity extremes, however, cause problems. High humidities lead to swelling, which may be disruptive, particularly for aged fibres. The opening of the polymer structure invites ingress by damaging pollutants and salts. The conditions also favour mould growth and microbiological breakdown.

Soaking in sea spray, of particular relevance to the *Victory* sail, will have led to increased swelling of the fibres, due to the penetration of hydrated sodium and chloride ions. On drying, swollen fibres will suffer dimensional changes, which in turn can lead to the distortion of the fabric. For the brine-laden sailcloth, further disruption of the linen would have ensued as the salt crystallised within the fibres.

Desiccation, under drier conditions (RH < 30%), will lead to increased cross-linking of the chains in the amorphous zones through secondary bonding, and brittleness. The overt results are shrinkage and loss of flexibility. As the temperature increases, adjacent hydroxyl groups on nearby chains will condense, leading to covalent cross-linking and further rigidity. The matrix components lignin and hemicellulose both moderate these processes – the former helps to resist swelling, due to its hydrophobicity and structural function, whereas the latter, with its branched chains and highly amorphous structure, will retain water.

There is another key thermally-promoted deterioration mechanism, oxidation. This is driven by free radicals and results in both cross-linking and chain scission, as well as discolouration to yellow/brown. These reactions will occur slowly at normal temperatures, but are accelerated by the presence of pre-existing free radicals, such as those arising from photolysis. Lignin and, to a lesser extent, hemicelluloses are both more susceptible to thermal damage than is cellulose.

Acid and Alkaline Hydrolysis.   Cellulose is susceptible to attack by a wide range of chemical agents, but primarily succumbs to either acid or alkaline hydrolysis. The damage starts in the accessible amorphous regions. The ready reaction with acid occurs randomly throughout the chain and leads to the scission of the glycosidic ether bond (Figure 9). The resulting fragments are termed hydrocellulose and though this material tends to be highly crystalline due to extensive hydrogen bonding after facile reorientation of the shorter chains, it is also mechanically weak and lacks the flexibility of native cellulose. Any source of acid is deleterious, with stronger acids, of course being the more potent, but even organic acids and inorganic Lewis acids will be effective.

The cellulose polymer exhibits a greater resistance to alkaline attack, which, unlike the random progress of acid hydrolysis, occurs only at the ends of

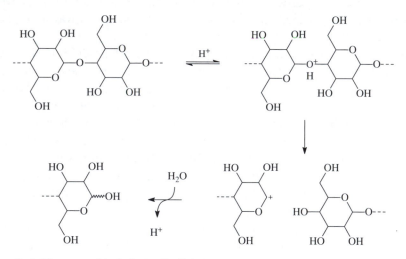

Figure 9 *Acid-promoted hydrolysis of cellulose*

the chains and proceeds via ring opening, reorganisation and β-elimination (Figure 10); the reaction will proceed along the chain in the direction of the exposed 'C_1' carbon. As this progressive depolymerisation ('peeling') occurs one unit at a time, it will not cause a significant deterioration in the physical properties of the fibre, unless there is pre-existing damage – the short chains of hydrocelluloses and oxycelluloses (the products of acid hydrolysis and photo-oxidation respectively) will degrade much more rapidly.

Photolytic Damage. Cellulose is susceptible to damage by light, especially radiation in the ultraviolet region. On exposure, the material undergoes various photodegradative reactions, including direct photolysis, photochemical and radical oxidation and photosensitised degradation (Figure 11a and 11b). These processes are accelerated by the presence of moisture and catalysts such as transition metal-based dyes and mordants. Extensive photodegradation will lead to depolymerisation and the formation of a variety of small, water-soluble acidic species, which tend to be yellow or brown in colour, so leading to discolouration. As a consequence of their solubility, they can readily be washed out of degraded fabrics, but this also risks further weakening.

Biological Degradation. Microorganisms which feed on cellulose excrete cellulase enzymes to break down the polymer. There are three types of cellulase: endo-β-1,4 glucanase breaks the β-1,4 bonds, exo-β-1,4 glucanase removes cellobiose units from the chain ends, and β-glucosidase cleaves the glycosidic link in cellobiose to form glucose. By-products of the further metabolism, such as hydrogen peroxide and organic acids, may cause additional

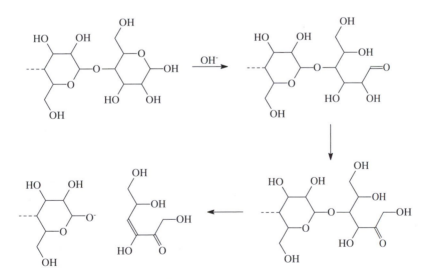

Figure 10 *Alkaline hydrolysis of cellulose*

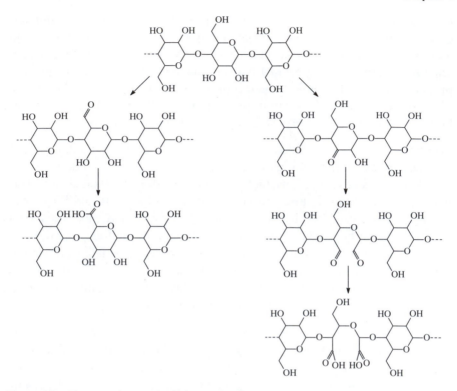

Figure 11a *Photo-oxidation of cellulose at pendant groups*

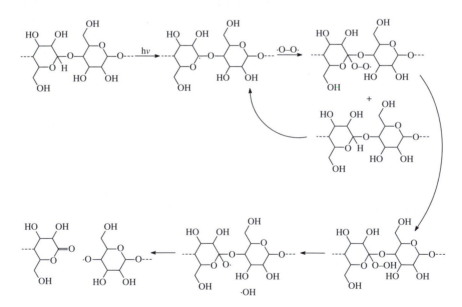

Figure 11b *Radical photolysis of cellulose*

damage to the polymer. As with other forms of attack, microbial damage will preferentially occur where there is improved access to the polymer chain, *i.e.* in the amorphous regions and in areas of pre-existing damage, such as photodegradation or swelling. Hemicelluloses and pectins are also the targets of biological degradation, but lignin appears to exhibit some resistance.

3.4 Acidity Measurements on the *Victory* Sail

In the absence of active biological attack, the two dominant processes leading to natural deterioration of cellulose are acid-promoted hydrolysis and oxidation. The latter is quite slow, and we would expect cellulosic artefacts that are above neutrality and kept in a stable environment to remain in relatively good condition for hundreds of years. However, those that are somewhat acidic can show significant signs of deterioration within a few decades.

As we have indicated above, the ageing of fibres like flax often results in the formation of acidic degradation products. These, together with adsorbed acidic atmospheric pollutants, can continue to promote further damage through the hydrolytic depolymerisation of cellulose. So, in relation to its stability, it seemed important to estimate the acid content of the *Victory* sailcloth. There are two approaches that can be taken: surface measurements on the fabric and acid extractions.

We recorded the surface pH by pressing a moistened non-bleeding indicator paper to the fabric for a short period, after which the resultant colour change was interpreted. These revealed pH values between 3.3 and 4.0 in the various locations tested, whereas for new linen sailcloth the value was around 6.0. While it is difficult to deduce the true acid content from this type of assessment, nevertheless values below pH 4 are generally regarded as a cause for some concern over the longer term, leading to recommendations for remedial action.

More detailed total acidity assessments were performed on short sections of the isolated yarns by extracting mobile and fibre-bound acid with degassed saline, before measuring the acidity with a microelectrode and pH meter. These saline extractions allow a better understanding of acid distribution, with the presence of salt ensuring the equilibration of immobile acid within the solution. Such acids may be strong (*e.g.* those arising from atmospheric pollutants) or weak (such as the carboxylic acids that are natural components of plant fibres or generated through oxidative deterioration). The total acid content deduced in this manner will represent a lower estimate since the weak acids may not be fully ionised, and this underestimate will be most marked for the more acidic samples.

Yarns from the modern, unaged sailcloth had a small acid content, probably just reflecting residual weak acid constituents of the plant fibres, such as hemicelluloses and lignin. The *Victory* sail samples had a higher acid content; for some yarns this increase was only marginal, while others had more than a 10-fold raised acid content, and in the case of a few heavily deteriorated specimens, a much-increased acidity was observed. These results contrast with the surface pH data, which suggested a similarly raised acidity across the whole sail. One interpretation of this anomaly is that in cases where the yarn acid content is lower, the acid is localised in the outer regions of the fibres, and that as the acid content rises it is found deeper within the fibres. This would be consistent with an initial surface oxidation, followed by the progressive deterioration of the secondary cell walls of the fibres. Titration implies that the majority of this acid is weak, derived from the oxidation of cellulose and other fibre constituents, which would fit with this hypothesis.

3.5   Conservation and Display of the *Victory* Fore Topsail

Before any treatment the sail was placed in an environmentally-controlled chamber and allowed to acclimatise under conditions of moderate humidity, where the fabric would be the most pliable. In the first phase, the conservators then surface cleaned the *Victory* sail with low suction vacuum cleaners and static cloths, removing a considerable quantity of particulate soiling. This not only improved the look of the sail but also reduced the risk of abrasive wear. Taking account of the need to handle the fabric gently and prevent undue stress, the sail was laid flat (over Tyvek®, an inert high density polyethylene sheet, on a suitable platform), and an inflatable roller was specially constructed to help to move the large artefact within a relatively confined space. There was then negligible risk of uneven short-term loading which could have ruptured the fabric.

Our mechanical tests suggested that it would be inadvisable to hang the sail from a yard, in the manner originally used onboard ship. Instead, it is more appropriate to display the artefact on a suitable frame that will support the weight of the fabric and prevent fibre slippage and permanent deformation. In terms of the sail's continued preservation, the apparent undue acidity of the fabric also needs to be borne in mind. Since, over the longer term, acid deterioration products can become mobile and may even be evolved in gaseous form, adjacent materials could become affected. This suggests that it is best to keep the sail unfolded and in an air conditioned environment to prevent any build-up of acidic gases, though for storage it might be appropriate to include acid sorbents between the layers of the cloth. Since some fibres are already quite acidic, deacidification also needs to be considered in the future. After weighing up the relative risks, non-intervention may be deemed the more appropriate course, avoiding the structural damage that might be caused by the deacidification process. More research is required to inform this key decision.

4 CASE STUDY 2: SILK FIBRES AND THE SHACKLETON ENSIGN

Sir Ernest Shackleton's white ensign, currently belonging to the Royal Yacht Squadron, presented a completely different set of challenges and problems. Having been involved in numerous expeditions in the first two decades of the 20th century, Shackleton's final voyage was planned as an expedition to the Antarctic, with the aim of circumnavigating the continent to make scientific and geographic observations. In September 1921 he set sail in the *Quest*, a wooden sealer refitted for the purpose, crossed the Atlantic to South America and then travelled southwards, coming within sight of the island of South Georgia on the 4th of January, 1922. Shackleton was taken ill in the early hours of the following day and died shortly after of heart failure. His body was sewn in canvas and covered with the white ensign, which as a member of the Royal Yacht Squadron he had the privilege of flying. Shackleton's body was to be returned to England, and was embalmed to this end, but at his wife's request he was instead buried on South Georgia. The ensign was presented to the Royal Yacht Squadron by Lord Shackleton, Sir Ernest Shackleton's son.

The ensign itself is constructed of red, white and blue ribbed-weave silk, with a pole sleeve of white, plain-weave cotton; it is approximately 211 × 117 cm in size (Figure 12). It was nailed to a wooden board, framed and glazed.

Figure 12 *The white ensign, before conservation treatment and after removing from its original frame*
(Photo: Mike Halliwell)

When it was received for conservation at the Textile Conservation Centre, it was in an extremely weak state, especially the areas of cream-white silk where significant sections of the fabric were missing and where there was some discolouration. There was extensive staining, principally from water and oil, as well as rust marks around the nail holes.

Again, an understanding of the chemistry and structure of the component fibres and their condition should inform the choices made for conservation treatment, display and storage. So, we continue this case study with such a description, before alluding to some of the original processing protocols, which could affect the silk fibres' stability. The condition of silk can be gauged at the microstructural and molecular level by contemporary analytical methods and we illustrate this, prior to discussing the details of silk deterioration. The particular rationale of our research is to inform conservation practice and we complete this section by referring to the specific treatment carried out on the ensign in light of its apparent condition.

4.1 The Chemical Composition and Microstructure of Silk

Silks are produced by a number of insect and spider species, but the major source of commercial textile silks are the silk moths of the family *Bombycidae*, of which the most important is the domesticated silkworm, *Bombyx mori*. The first known references to the culture of silkworms (sericulture) come from China in the second or third millennium BC, from where it spread through Asia, then on to the Middle East and eventually Europe.

Moth larvae produce the polymeric material from a liquid crystalline phase to form a protective cocoon; silk is extruded from the spinneret as a 'bave', consisting of two roughly triangular fibroin filaments ('brins'), about 20 μm across, bound together by a second protein, sericin (Figure 13). In commercial production, the chrysalid is killed before it can hatch (which would damage the thread), and a continuous filament can then be unreeled from each cocoon, giving typically 500–800 m of useable thread.

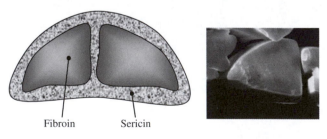

Fibroin   Sericin

Figure 13 *Cross-sections of silk fibres: (a) cartoon of a native silk fibre and (b) scanning electron micrograph of a fractured, processed textile fibre (10 μm across)*

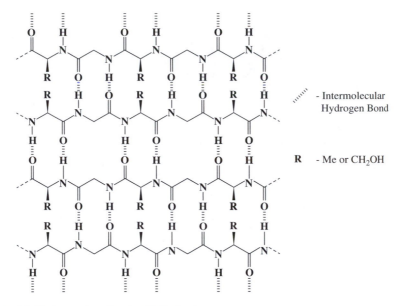

Figure 14 *Hydrogen bonding in silk β-sheets*

Fibroin is largely composed of three amino acids, glycine (45%), alanine (29%) and serine (12%); it can be divided into two main components, the crystalline and the amorphous regions, which are in the approximate ratio 3:2. The crystalline regions are formed of a hexapeptide motif (shown below), folded into an anti-parallel β-sheet secondary structure, held together be extensive hydrogen bonding (Figure 14) and strongly aligned with the fibre axis.

-Gly-Ala-Gly-Ala-Gly-Ser-

or

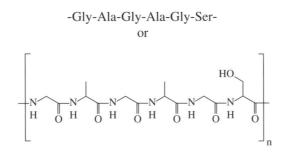

The crystalline close packing of this region is facilitated by the small, non-bulky side-chains of the residues (—H, —Me or —CH$_2$OH); see Figure 15. The unit cell is built from four protein chains, with *ab* being in the plane of the β-sheets and *b* lying along the fibre axis.

The highly crystalline nature of fibroin leads to the characteristic properties of silk fibres: a good mechanical strength and limited extensibility, arising from the extensively inter-bonded, fully extended nature of the protein chains, and

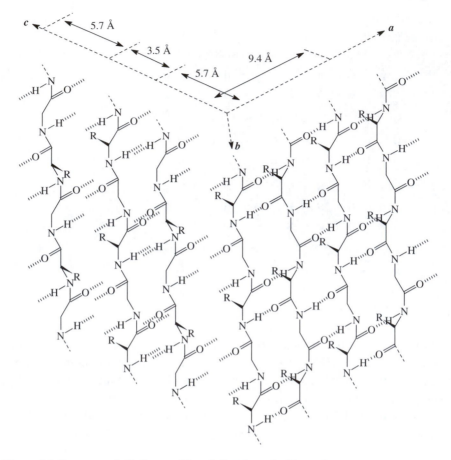

Figure 15 *Structure of silk β-crystallites; b lies along the fibre axis*

a high degree of chemical resistivity. The amorphous and semi-amorphous regions that comprise the remainder of the fibre contain a high proportion of residues possessing bulky side-groups, which dictates the low degree of ordering here.

As with other natural fibres, silk has a hierarchical microstructure – about five anti-parallel β-sheets, each with around 12 chains, aggregate to form parallel, crystalline microfibrils (approximately 10 nm in diameter), bundles of which make up fibrillar elements (roughly 1 μm across), which in turn associate to comprise the individual fibroin filaments (7–12 μm); at each level of organisation, the ordered elements are embedded within amorphous matrices derived from the non-crystalline components. Once again, then, the behaviour of the structural composite can be understood in terms of the semi-crystalline array of its component parts.

Sericin, the protein that binds the pairs of fibroin filaments as they emerge from the silkworm, and which may have a role in dehydrating the fibroin and encouraging its crystallisation, has a markedly different composition and structure to that of fibroin. It is largely amorphous and is rich in serine ($\sim$32%), aspartic acid ($\sim$14%) and glycine ($\sim$13%); there is a much greater proportion of residues with polar and/or bulky side-chains. The predominance of these polar, hydrophilic groups means that sericin is readily soluble in hot water.

4.2 Silk Processing for Use as a Textile Fibre

Silk yarns are thrown or spun; the former are of higher quality being produced from bundles of lightly twisted continuous filaments, while the latter is spun from shorter, broken sections of fibre. The fibre can be used 'raw', with the sericin intact, or 'degummed', when the sericin has been removed to yield individual fibroin strands. Degumming is usually achieved by the use of boiling water or steam, by immersion in an alkali (soap) bath, by fermentation (microbial or enzymatic breakdown of the protein), or, more recently, by the use of acids. The degummed fibres are smoother and more lustrous, but their physical strength is often reduced and they will absorb less water (which in turn affects the affinity for dyes and weighting agents).

Following degumming, silk can be bleached to remove any residual colour. Historically sulfur bleaching ('stoving') was common, until superseded by modern 'wet' chemical methods, generally including hydrogen peroxide, although sometimes sodium peroxide, various bisulfites and '*blueing*' agents.

A particularly important aspect of silk processing is 'weighting'. Historically, silk was sold on the basis of weight, but degumming can reduce this by as much as 25%, so methods were sought to restore this lost weight. Subsequently, weighting was used to fraudulently increase the mass of the fibre beyond its original level, and finally it became an accepted method of preparation to improve the feel (crisper and stiffer) and drape of the fabric. In extreme cases, silks may be loaded by up to 400%, though this can lead to rapid deterioration of the fabric.

A variety of different weighting agents and methods have been used since the middle ages, including gum Arabic, salts of tannic, gallic and formic acids, catechu and logwood ('vegetable weighting'), protein glues, waxes, sugar and various metal salts. It is this last category, the metallic salts, that have found the most widespread use, and many serve a dual purpose, also acting as mordants especially for darker dyes. They include compounds of iron, lead, tin, aluminium and zinc, and more rarely arsenic, barium, bismuth, chromium, copper, magnesium and tungsten; the most common of these are salts of lead and tin (in the form of lead(II) acetate and tin(IV) chloride). These agents are readily absorbed within the amorphous regions of the fibre where they tend to concentrate. All weighting agents, and especially the metal salts, can present serious problems in terms of silk preservation as many of them act to catalyse degradation reactions. In addition, the processes used to apply the materials can also lead to damage (*e.g.* the use of acidic baths). Consequently, many historic silks, especially 19th and early 20th century European silk artefacts, are in a relatively poor state.

4.3   The Condition of the Shackleton Ensign

Once released from its frame, the true condition of the ensign was revealed. Immediately apparent were the vivid colours of those areas that had been protected from light compared with the faded tones of the exposed regions. The extreme fragility of the fabric was evident. There were splits where the flag had been folded, as well as numerous smaller fractures and several areas of complete loss (Figure 12). Closer examination highlighted the highly degraded nature of the silk fibres, characterised by thinning, brittleness and various patterns of breakage. The majority of the damage was concentrated in the areas of the cream-white silk; bleaching or finishing treatments applied to this silk at the time of manufacture could have directly damaged the fibres or rendered them the more susceptible.

To test this hypothesis in part, we subjected individual loose threads from the ensign to X-ray microanalysis in a scanning electron microscope. Characteristic X-rays are emitted from the constituent atoms of a specimen as a result of electron-beam induced core electron ionisation processes. The X-ray spectrum resulting from the cream-white silk threads indicated quite heavy tin weighting (Figure 16); the coincidental presence of silicon and phosphorus suggests

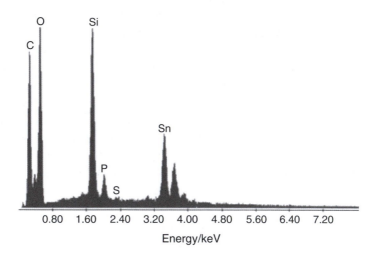

Figure 16 *The X-ray spectrum generated by a cream-white silk thread from the Shackleton ensign when probed in a scanning electron microscope*

original use of the 'dynamite' weighting process. The red weft threads proved to be just lightly tin weighted, while the red warp threads and blue silk threads were devoid of inorganic weighting. It seems reasonable to ascribe the friable nature of the cream-white silk to either direct or indirect consequences of the original tin weighting process.

4.4   Assessing the Condition of Silk by Microanalytical Techniques

It was clear from the outset that the majority of the ensign silk had suffered significant deterioration and was mechanically weak and brittle. In other circumstances, more detailed studies which will define the physical and chemical state of better-preserved materials, are of value in informing conservation decisions. Practical and ethical considerations dictate the application of analytical methods which require only minute specimens consisting of no more than a few fibres a millimetre or two in length. Spectroscopic, X-ray diffraction and mass spectrometric techniques, which can report on changes at the microstructural and molecular levels, are the subject of promising current research. Viscometric and chromatographic procedures which reveal the relative fibroin polymer chain length are already routine. The most useful methods will show unique signatures for degraded material that correlate with the physical properties of the aged threads. For example, we have investigated a variety of silks by high performance size exclusion chromatography and demonstrated a simple relationship between the strength of the silk and the fibroin peak retention time, which relates to the polymer molecular weight (Figure 17).

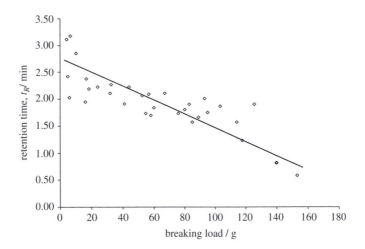

Figure 17 *The breaking load of a variety of silk samples plotted against their retention time, t_R, in high performance size-exclusion chromatography*

For this experiment, small fibres were immersed in a concentrated lithium thiocyanate solution. This disrupts the hydrogen bonds that hold the protein chains together, allowing the fibroin polymer to dissolve, but without cleaving the chains. The sample solutions were loaded on to a size-exclusion column, which is able to separate molecules on the basis of their effective length. Smaller polymer chains are washed through more slowly. The degraded silk samples had a much greater retention time (the time that the polymer fragments stay on the column) than pristine silk, demonstrating the fragmentation of the polymer chains on ageing.

4.5   Degradation of Silk Fibres

To generate deteriorated fibres for the above experiment, new silk was either thermally or photo-degraded. Silk is particularly susceptible to light damage and it is suspected that this was a significant factor in the case of the Shackleton ensign, particularly for the tin weighted cream-white threads.

As for linen and other natural fibres, silk is sensitive to a variety of environmentally driven degradative processes, though in most cases the actual damage is caused by hydrolysis and/or oxidation. Attack on the polymer chains is generally initiated in the amorphous zones as a consequence of their more open structure and the incidence of reactive amino-acids (specifically histidine, lysine, phenylalanine, proline, threonine, tryptophan, tyrosine and valine).

Humidity and Heat.   Water bonds strongly to silk due to the high proportion of polar groups in the material, particularly within the amorphous regions. At 65% humidity, bound water can represent nearly 10% of the weight of the fibre, leading to some swelling. The presence of salts will alter the pattern of swelling, as will pre-existing degradation – in heavily deteriorated samples this can result in the catastrophic breakdown of the fibre structure.

Changes in the environmental conditions during its previous display and storage seemed to have affected the Shackleton ensign. The evident distortion of the edges of the fabric and the apparent stresses introduced around the nails used to attach it to the board suggested to us that there had been moisture-induced dimensional changes after mounting.

Because water acts as a plasticiser and there is a strong affinity between the fibre and water, silk retains its flexibility at humidities as low as 40%. Below this value, desiccation occurs resulting in brittleness and rigidity; the dimensions of the unit cell reduce as well. Besides encouraging loss of water, elevated temperatures will also result in free-radical thermal oxidation, occurring primarily at amino acid residues with side-groups that readily lose hydrogen (Figure 18). In addition, the fibres will undergo yellowing as additional chromophores are produced.

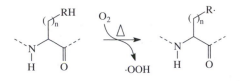

Figure 18 *Free radical thermal oxidation of silk*

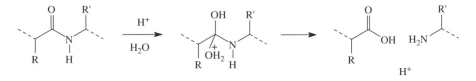

Figure 19 *Acid hydrolysis of silk*

Acid and Alkaline Hydrolysis. Silk is quite resistant to chemical attack, as a result of its highly crystalline nature. However, some reagents, *e.g.* those used in the original processing of the cloth and atmospheric pollutants, will affect deterioration, notably over the longer term.

Acids, particularly in higher concentrations, attack the amorphous regions relatively rapidly and at random points along the chain, leading to the hydrolysis of peptide bonds (Figure 19). Microstructural changes will also ensue as the hydrogen bonding and salt linkages that dictate the secondary and tertiary structure of the polymer are disrupted. This results in brittleness and the loss of mechanical strength. Additionally, the presence of acids can induce a rearrangement of peptides to esters and ethers. The action of some acids is used commercially to induce sought-after properties: weak acid (particularly acetic and tartaric) treatment imparts 'scroop', a characteristic rustling believed to be the result of microcrystalline deposits, while concentrated sulfuric acid, which leads to the contraction of the fibres and the modification of side-chains, alters lustre and softness to yield fibres for the production of silk crêpe.

Alkalis hydrolyse silk more slowly than acids, the reaction proceeding most rapidly from the ends of the chains. However, as with acids, the presence of alkaline agents can additionally break hydrogen bonding and salt linkages, leading to a similar disordering of the structure. Concentrated alkaline solutions, especially if hot, then cause the complete dissolution of silk. Although generally effecting chain shortening (Figure 20), alkaline conditions also have the potential to cause cross-linking, *e.g.* via the formation of lysine to serine bridges (giving lysinoalanine), leading to some increased rigidity.

Fibre-included weighting agents can promote hydrolysis of the fibroin – *e.g.* over time tin salts are themselves hydrolysed to amphoteric species that can affect both acid and alkaline attack.

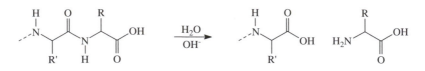

Figure 20 *Alkaline hydrolysis of silk*

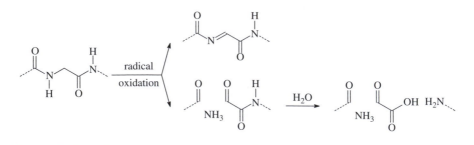

Figure 21 *The radical oxidation and scission of the peptide bond, following photodegradation*

Besides acids and alkalis, strong oxidants used to bleach silk also have the capacity to seriously degrade the fibres. The most significant reactions occur at tyrosine and threonine side-chains, resulting in their oxidation to acids and the breaking of surrounding peptide bonds; cross-links are also generated, *e.g.* between lysine and tyrosine residues.

Photolytic Damage. Silk is particularly susceptible to damage by radiation, especially in the ultraviolet region, typified by weakening and embrittlement, accompanied by yellowing. The explanation for this lies in the relatively high proportion of aromatic residues in the protein, all of which are prone to photo-oxidation. The degradation products are generally not well characterised, but include chromophoric derivatives of tryptophan and tyrosine. Besides side-chain modification, the radicals generated will affect cleavage of neighbouring peptide bonds, weakening the fibres (Figure 21). A variety of other photochemical reactions can also occur, including cross-linking in the amorphous regions, leading to a loss of flexibility, primarily due to the reactions of photo-activated tyrosine residues with each other or with lysine. The presence of certain dyes, such as anthraquinones, will accelerate these various processes, and moisture also has an influence.

Some weighting agents will act to accelerate the photo-deterioration of silks too. Metal ions, such as tin, adopt catalytic rôles in degradation reactions, whereas tannin compounds can reduce the pH of the silk to the point where it becomes particularly susceptible to photodegradation.

Biological Degradation. Microbial (enzymatic) attack can occur on silk, but it is the water-soluble sericin, that is the more readily depolymerised.

Again due to the highly crystalline nature of fibroin, degummed silks are relatively resistant to damage from micro-organisms. A consequent advantage is that proteolytic enzymes can be used in the degumming process.

4.6 Conservation Treatment of the Shackleton Ensign

The severely degraded nature of the silk, particularly the cream-white threads, was immediately obvious on examination of the ensign. Even very careful handling caused additional damage to the stiff, fragile fabric; this proved to be a particular problem in areas of marked staining. Our analytical work had suggested a reason for this, the heavy tin weighting of the cream-white silk. In the 'dynamite' weighting process the fabric is immersed successively for many hours in solutions with pH values as low as 1 and as high as 10. The immediate damage caused by this and any prior bleaching will have sensitised the silk. Subsequent light-promoted deterioration, probably accelerated by the suffused tin, is likely to have been responsible in main part for the catastrophic ageing that has ensued, though other agents will have been active. Tin oxohydroxide, formed from the deposited tin salts, will have effected hydrolytic cleavage of the fibroin. There was also evidence that the deterioration of the original wooden frame had resulted in the acidification of the adjacent textile, leading to further damage to the underlying silk fibres.

In these circumstances, ideally, measures should be taken to stabilise such a material, inhibiting further decay and consolidating the fragile fabric so that the hazard of handling is mitigated. Initially the ensign was gently surface cleaned with a low suction vacuum cleaner, to remove loose debris such as broken fibres and dust. However, the risk to the ensign of removing ingrained soiling and relieving acidity by more interventive approaches was considered too high.

Tests confirmed that the cloth was too frail for wet cleaning and deacidification, *i.e.* soaking with a buffered weak detergent solution. Sample fibres immersed in water became considerably swollen and soon fragmented. This damage was exacerbated when samples were moved, as would be necessary during cleaning, rinsing and drying. (Where exposure to water is inappropriate, textiles may still be treated with an organic solvent to remove less polar soiling. However, dimensional changes are again likely and furthermore the solvent may serve to desiccate the fibres.)

Simple humidification, which is often used to increase the pliability of fragile textiles for handling and so facilitate stitching, was also thought to be too risky. Nonetheless, consolidation of the brittle fabric and a supporting treatment was obviously necessary to preserve the integrity of the ensign, such as it was. Consequently, a suitably dyed silk crepeline net was adhered to the ensign with a water-dispersed butyl methacrylate adhesive. Of course, contact between the silk and water had to be avoided, so the net was first coated with the adhesive, which was allowed to dry, and the crepeline then laid over the ensign and the acrylate reactivated with a heated spatula to effect bonding. The ensign was finally remounted on a rigid, chemically-stable board covered with conservation-quality materials: polyester felt, downproof cotton and a layer of cotton lawn dyed to match the colours of the flag.

While it was not possible to remove or neutralise the intrinsic effectors of ageing, nonetheless recommendations could be made concerning the lighting during display and the relative humidity of the storage environment which would now ensure that the ageing rate was minimised.

5   CASE STUDY 3: WOOL FIBRES AND THE TREE OF JESSE TAPESTRY

The *Tree of Jesse* (Figure 22) is a 15th century Rhenish ecclesiastical tapestry, comprising two parts, an altar frontal and a smaller superfrontal; it currently belongs to the Whitworth Art Gallery, Manchester.

The frontal itself depicts the biblical Tree of Jesse, a popular mediaeval subject illustrating the ancestors and prophets of Christ; branches spring from the seated figure of Jesse, forming scrolls which frame the 12 Kings of Israel, all wearing 15th century aristocratic dress, above whom in the central position are the Virgin and Child. The piece is woven with linen warps, and wefts, principally of wool, along with silk, metal threads and linen; inscriptions are worked in ink, as well as laid and couched embroidery, and facial details are painted on to the silk wefts. The precision and delicacy of the weaving suggest that it was the work of a major workshop. The tapestry is rectangular and measures 2068 mm in width and 649 mm in height.

The accompanying superfrontal is a narrower piece (1995 mm wide and 163 mm high) with the legend 'PUER NATUS EST NOBIS ET FILIUS

Figure 22 *The Tree of Jesse Tapestry*
(Photo: Mike Halliwell)

DATUS EST NOBIS' (The Father is noble and so is the Son) in Gothic lettering, at the centre of which are Mary, Joseph and the infant Jesus in the stable, along with Sts Catherine and Barbara. Like the frontal, the warps are linen, with wool the predominant component in the weft, alongside silk and metal threads. The tapestry was decorated with a silk fringe and metal thread braid edging, though investigations revealed that this latter adornment was a later addition, as it contained brass filaments manufactured by a method not developed until the 18th century. Although they are both of a similar period and origin, it is believed that the superfrontal was not originally intended as a companion piece for the frontal, but that the two were paired at some time after their manufacture.

When it was received for conservation, the tapestry was soiled in places with a light grey dust, as well as fibre dust along the lower edge of the underside, and there was some staining around the Virgin and Child. Pink deposits were observed on the St Lucas roundel and waxy deposits near the scroll of Ahaziah. Some of the exposed linen warps were discoloured to brown; many of the wefts have faded. There were holed areas where both the linen warps and wool wefts had deteriorated; the beige-coloured wool appeared to be in particularly poor condition, and was largely absent. Many of the metal threads exhibited signs of corrosion. Evidence of previous repairs was found, in the form of small linen patches used to reinforce weakened areas, and coarse stitching around some of the figures.

5.1 The Chemical Composition and Microstructure of Wool

Wools and other similar mammalian hairs are largely composed of keratin proteins. However, unlike the other natural proteinaceous fibre, silk, wool is cellular in nature: the fibres consist of relatively hard, flattened, overlapping cuticle cells, which surround the central cortical cells; in some fibres, these may in turn surround a hollow medulla (Figure 23).

The cuticle can vary in thickness between one cell, for the finest wools, up to several dozen. These cells contain amorphous keratin arranged in layers, with the resistant exocuticle overlying the inner endocuticle. The resilience of the exocuticle derives in large part from its sulfur-rich (cysteine) nature, which allows the formation of the disulfide bridges that reinforce the polymer; the endocuticle has much lower sulfur content and so is more susceptible to deterioration. The exposed surface of the cells is covered by a protein membrane, the epicuticle, along with waxes and fatty acids, which impart a further degree of chemical and mechanical resistance. It is the overlapping cuticle cells that give wool fibres their characteristic scaly appearance (Figure 24). Lying between the cuticle and the cortex is a continuous band of intercellular materials, which, in combination with the two adjacent cell walls, lends a resistance to penetration by water and other chemicals.

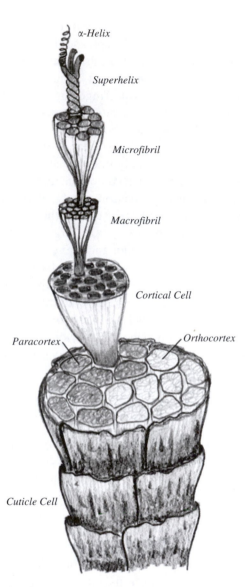

α-Helix

Superhelix

Microfibril

Macrofibril

Cortical Cell

Paracortex

Orthocortex

Cuticle Cell

Figure 23 *The hierarchical microstructure of wool*

The bulk of the fibre consists of the cortex. The elongated cortical cells are composed of crystalline proteins within a sulfur-rich amorphous protein matrix. The keratin adopts two principal secondary structural motifs, the α- and γ-helices, which form due to extensive hydrogen bonding and steric factors, with large side-groups arranged around the spiralling peptide backbone. In the crystalline regions of the fibre, superhelices or protofibrils are formed from the α-helices (Figure 25) which are themselves arranged in a helical fashion. These are then organised to form microfibrils, which in turn form macrofibrils; these,

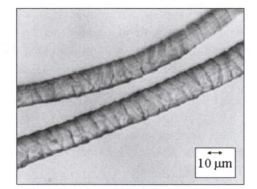

Figure 24 *Typical wool fibres*

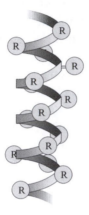

Figure 25 *Schematic of an α-helix*

located within the amorphous bulk of the keratin, comprise the proteinaceous regions of the cortical cells. The amorphous matrix itself is principally stabilised by the large number of disulfide bridges, which link proximate cysteine residues. The cortex can be further divided into two general regions, the orthocortex and the paracortex, the former containing more crystalline fibrils and the latter more largely composed of amorphous material.

Finally, depending on the species from which the wool is taken, a medulla – a hollow, air-filled cavity – may be present at the centre of the fibre. This feature is generally found only in coarser fibres, and may take a number of different forms, ranging from a continuous unbroken channel (*e.g.* many sheep's wool fibres) to interrupted, lattice-like structures (*e.g.* rabbit fur).

5.2 Wool as a Textile Fibre

Wool lacks the mechanical strength of silk, due to its cellular nature and lesser degree of crystallinity. However, these factors also give it a greater extensibility and elasticity. Wool is often degreased to remove the natural oils (lanolin in the case of sheep's wool) that coat it and give it water-resistant properties.

Wool is particularly valued as a textile fibre for its desirable properties in terms of comfort. When it absorbs water, heat is liberated, making woollen fabrics ideal for cold, wet conditions; this ability to trap and retain water also keeps the relative humidity at the skin comparatively high, making it a suitable material for hot, dry environments. However, wetting weakens the fibres and may promote bulk dimensional changes in fabrics. The bulky, crimped nature of wool fibres ensures that the textiles contain a high proportion of air space, giving them good insulative properties.

Wool fibres may be subjected to *fullering* – the fibres are treated with an appropriate agent (traditionally mercury compounds were used), which causes the individual scales on the fibre to stand proud of the surface, causing adjacent fibres to become entangled, and increasing the bulk of the material. The process can be employed to make non-woven woollen textiles (*i.e.* felt), but is also employed to modify the properties of fibres used for yarns and fabrics, and can give a durability to wool fabrics that is not found in the individual fibres.

Wool and hair fibres can be taken from a wide range of animal species such as sheep, goats, camels, yak, musk ox, llamas and related species, and 'fur' animals (rabbits, foxes, beavers and the like), among others. The properties of these fibres are dictated by the source, and these variations can influence length, diameter and fineness, mechanical characteristics and durability, colour and dyeability.

5.3   Degradation of Wool

Like other natural fibres, wools are subject to attack by heat, light, acid and alkaline hydrolysis, the action of fungi and other microorganisms, and by mechanical damage (wear and tear), but in contrast to the other fibres, wool is also particularly prone to insect attack. As wool is primarily a proteinaceous fibre, the chemistry of wool is largely dictated by protein chemistry. In general these fibres are more susceptible to chemical deterioration than are silks, due to differences in the chemistries, structures and micro-structure of the fibres. The composition of silk fibroin ensures that it can adopt extensive β-sheet structures, which in turn form chemically-resilient crystallites. Wool, on the other hand, is significantly more amorphous in structure, allowing degradative reactions to propagate much more rapidly, particularly once the resilient outer layers of the fibre cells have been damaged. Oxidation will attack the protein chain in general, especially at susceptible side-chains, and most notably will lead to the scission of the structurally-important disulfide bridges; this oxidised wool is soluble in basic solutions. However, wool fibres tend to exhibit a greater resistance to photodegradation than do silks.

As the formation of disulfide bridges between proximate cysteine residues plays a particularly important rôle in physical properties of wool, any reagents or conditions that interfere with these bonds will have a significant effect on the fibres. A particular and related problem associated with the deterioration of wool fibres is the release of volatile sulfur compounds, which may then attack adjacent materials: many of the silver-containing metal threads found on the *Tree of Jesse* tapestry show signs of surface corrosion, in the form of silver sulfide.

Heat and Humidity.   Wool will normally bind about 16–18% moisture, but on soaking the fibres can bind up to 200% of their dry weight in water; as a result, degraded fibres are particularly susceptible to damaging swelling, and are also much more prone to mechanical damage in this state. Swelling and desiccation may also lead to the rearrangement and formation of new bonds, leading to the modification of physical properties and dimensional changes in the bulk fabric. The presence of moisture, particularly in warm conditions, will also serve to promote other deleterious reactions.

Acid and Alkaline Hydrolysis.   The protein hydrolysis of wool, when faced with acidic or alkaline conditions is similar to that of silk, as noted above. As with silk, the presence of acidic or alkaline conditions will lead to the hydrolysis of peptide bonds, although of the two fibres, wool will generally degrade the more rapidly, particularly if the resilient outer layers are damaged, as a result of its lesser degree of crystallinity. Due to the greater number of cysteine residues, wool proteins are also more prone to cross-linking through the formation of lanthionine and lysinoalanine bonds. A further effect of acid or alkaline treatment is the degradation and loss of the protective waxes and fatty acids, which coat the fibres, facilitating further attack.

Photolytic Damage.   Exposure to light, particularly below 380 nm and in the presence of water, will initially cause wool fibre to undergo yellowing, due to reactions within the amio acid side-chains, and eventually will lead to chemical modification, mechanical weakness, embrittlement and the loss of flexibility due to various free radical reactions; photo-oxidation can lead to the scission of disulfide bonds. The histidine, tryptophan, tyrosine, methionine and cysteine residues are particularly susceptible. Longer wavelengths may cause bleaching. However, the fibres do exhibit a greater resistance to photodegradation than does silk.

Biological Degradation.   Wool may be targeted by a variety of keratinophilic bacteria and fungi, which break down the component proteins via enzymatic oxidation, reduction or hydrolysis; this type of attack predominantly tends to occur in the amorphous regions, particularly in the cuticle. Wool fibres are also susceptible to attack by several species of moths and beetles.

5.4   Assessing the Condition of Wool by Microanalytical Techniques

Electron microscopy can provide valuable information about morphological changes in the fibres, and can potentially give an indication of the nature and cause of degradation, such as photolytic damage, swelling, desiccation or abrasion. The appearance of fracture surfaces is a particularly useful source of information.

Amino acid analysis is also potentially of value in monitoring and assessing deterioration. By following the way in which the proportions of various residues vary over time, it is possible to gain an understanding of the type of damage that has occurred to the fibres: *e.g.* the cystine disulfide bridge is particularly susceptible to photo-oxidation, and this reaction is accompanied by a corresponding increase in cysteic acid, a process which can be followed using this type of analysis.

Chromatographic techniques may also reveal a range of important data. End-group analysis will give an indication of the breakdown of the peptide chain, and the measurement of extracted soluble protein will similarly indicate the extent to which the polymer has deteriorated. These techniques, alongside mass spectroscopy, can also reveal the presence of dyes and other treatments.

5.5   Conservation and Treatment of the Tree of Jesse Tapestry

An approach of minimal intervention was taken for the conservation treatment and subsequent mounting of the tapestry. Gentle vacuum cleaning was employed to remove the particulate soiling, which, if left in place, may have caused abrasion. Those earlier repairs and alterations that were in good condition were left intact; only those which were found to be unstable were treated. The calico lining, which joined the frontal to the superfrontal, was removed, as this was not thought to have any historical significance, and this enabled different support methods to be employed for the two pieces, suitable to their individual requirements – the Nativity scene on the superfrontal, for example was reinforced with a linen patch. Both sections were then stitched to conservation-quality boards to provide protection and long-term stability, and to limit the necessity of further intervention. Pest monitoring was recommended as a long-term strategy, due to the particular susceptibility of wool fibres to attack by insects.

6   CONCLUSIONS

We began this chapter by suggesting that a conservation scientist would need a thorough appreciation of the underlying science to best advise on the long-term preservation of historic artefacts like the *Victory* sail, Shackleton's ensign and the *Jesse* tapestry. In the associated investigation carried out to support the textile conservator, the constituent materials of the artefact have to be characterised to give an appreciation of their current physical and chemical state.

The conservator needs to be able to understand whether the preservation of the artefact is compromised in any way by its present condition, by what means the materials can be stabilised and how they will behave in the future. If the knowledge to provide this understanding is not already in the public domain, the conservation scientist will need to investigate the nature of the materials and especially their deterioration as brought about by heat, light and moisture, and perhaps gaseous pollutants and microorganisms.

We intended our three case studies to not only provide suitable vehicles for discussing the nature of our selected natural fibres, but also to illustrate the vital role for science and the conservation scientist in the continuing preservation of our cultural heritage.

ACKNOWLEDGEMENTS

We are particularly grateful to Lt Cdr Frank Nowosielski, Commanding Officer, *HMS Victory*, the Royal Yacht Squadron and the Whitworth Art Gallery for allowing us to present the case studies.

We would also like to acknowledge the valuable support of the following during our research on the *Victory* sail: the Society for Nautical Research, Colin Appleyard (Hood Sailmakers), Kate Gill (Textile Conservation Centre), Peter Goodwin (Keeper, *HMS Victory*) and Mark Jones (the Mary Rose Trust).

As ever, we owe much to our colleagues at the Textile Conservation Centre, Winchester, especially the Director, Nell Hoare.

REFERENCES AND FURTHER READING

We offer the following suggestions for those interested in learning about other textile fibres and reading more about the science of textile conservation.

V. Ashton, A stitch in time, *Chem. Brit.*, 2001, **37**(10), 46–48.

F.G. France, Scientific analysis in the identification of textile materials, in *First Annual Conference of the AHRC Research Centre for Textile Conservation and Textile Studies, Scientific Analysis of Ancient and Historic Textiles: Informing Preservation, Display and Interpretation, Winchester, 13–15 July 2004*, R. Janaway and P. Wyeth (eds), Archetype, London, 2005, 137–142.

K.L. Hatch, *Textile Science*, West Publishing Co., Minneapolis/St Paul, 1993.

M. Lewin and E.M. Pearce (eds), *Handbook of Fiber Chemistry*, 2nd edn, Marcel Dekker, New York, 1998.

Á. Tímár-Balázsy and D. Eastop, *Chemical Principles of Textile Conservation*, Butterworth-Heinemann, Oxford, 1998.

CHAPTER 5

Leather

ROY THOMSON

Consultant in Leather Science and Conservation, Oundle, Peterborough PE8 4EJ, UK

1   INTRODUCTION

It has been suggested that the hides and skins of animals, killed for food were used for clothing, shelter and other purposes by our hominid ancestors some 2 million years ago, long before the evolution of *Homo sapiens*. The fact that Neanderthal Man not only survived during the last ice age, about 100,000 years ago, but expanded during that period into the bleak tundra regions of Northern Europe and Central Asia indicates that the dressing of skins for the production of warm, protective clothing and tent coverings had been developed to a high order. Evidence of the arrival of Cro Magnon Man in Europe 35,000–40,000 years ago, includes the discovery of many types of skin-working tools such as scrapers, burnishers, knives and awls made from bone or stone. Many are of such delicate construction that it can be surmised that the art of producing soft, flexible leathers had been completely mastered. Indeed the types and range of leather-working tools uncovered suggest that in some regions, 20,000 years ago, leather-making techniques were at least as advanced as those employed during the nineteenth century by Native Americans and peoples from Central Asia. It is not surprising, therefore, that the production of leather has been described as "man's first manufacturing process".

2   THE NATURE AND PROPERTIES OF LEATHER

The skin of an animal can be transformed into a variety of products having a wide range of characteristics. Leather and other skin-based materials can be hard or soft, flexible or rigid, stiff or supple, thick or thin, limp or springy. A leather ideal for glove-making for instance would not be suitable at all for shoe soling. The properties of any particular material will depend both on the nature of the skin chosen and on the processes employed. It has been the skill of the tanner throughout the ages to make a product with just the combination of properties required by the end user.

An examination through the thickness of a skin reveals that it consists primarily of long fibres and fibre bundles interwoven in three dimensions within a jelly-like ground substance. Other features are present, which are vital for the functioning of the skin in life but which are generally removed during processing. These include a keratinous epidermal layer in which hairs and hair root structures are embedded, and muscles, blood vessels and fat cells. It is, however, the intricate three-dimensional woven fibrous structure which predominates, and it is this that imparts many of the unique physical properties characteristic of the wide variety of materials that can be made from skin. These include a relatively high tensile strength with particular resistance to shock loads; low bulk density; flexibility where required; resistance to abrasion, tearing and puncturing; good heat insulation and an ability to be stretched and compressed without distorting the surface.

2.1 Criteria which Define Tannage

The above physical characteristics are common to most skin-based products. Linguistic studies show, however, that even from early times, materials such as rawhide, oil-tanned pelts, parchment, alum-tawed skins and vegetable-tanned leathers were distinguished from each other. The question arose as to what are true leathers and how they can be differentiated from other materials made from skin without undergoing a tanning process. A number of criteria have been suggested. These include resistance to microbiological decay, physical properties such as flexibility and a "leathery handle" and elevated shrinkage temperatures.

Resistance to Microbiological Attack. Under normal ambient temperatures, a wet raw skin will decay rapidly due, primarily, to the action of bacterial proteolytic enzymes. Leather, on the other hand, resists such microbiological attack even if it remains wet. This fundamental difference has been used to define tannage.

There are, however, a number of techniques employed in the leather-making industry to inhibit bacterial action and prevent the degradation of freshly flayed pelts. These include drying the skin, salt curing and acid pickling. These enable raw materials to be transported from source to production unit, and to be held temporarily in a safe condition until they are required for processing. This resistance to decay is, however, lost if the skins are wetted. In a similar manner, alum-tawed pelts and parchment, both renowned for their longevity, degrade rapidly if they become wet for any appreciable time.

Many indigenous peoples treat skins by impregnating them with fats. They are then allowed to dry under controlled conditions while being worked mechanically. This procedure both coats the individual fibres and fills the spaces between them with the fatty material. This renders the skins water-resistant and even if they are subjected to wet conditions, the fibres themselves remain too dry for bacterial action to take place. These materials therefore appear to satisfy the criteria of resistance to microbiological attack. Such products are found widely in ethnographic collections and have been called pseudo-leathers.

These should not be confused with oil-tanned skins such as chamois washleathers or the buff leathers employed widely by seventeenth century armies to make protective jerkins. These are not impregnated with stable, water-repellent fats but treated with reactive, oxidisible oils, which undergo chemical reactions with the skin during processing to give a proteolytic enzyme-resistant product.

It can be seen that methods can be employed to produce materials, which are apparently resistant to microbiological attack, but which are not truly leathers.

Physical Properties.   It is generally considered that if a raw skin is allowed to dry it will become hard, horny, translucent and relatively inflexible. If the hair has been removed first, these characteristics have been exploited to produce such diverse objects as rawhide mallet heads and dog chews. Leather on the other hand is expected to dry to give a soft, flexible, opaque product with a characteristic feel. This has been cited as evidence that tannage has occurred.

It is true that if the skin is simply dried in an uncontrolled manner the result is likely to be as described. If, however, it is impregnated with fats as in the production of pseudo-leather the product will have all the physical characteristics of a properly tanned skin.

While parchment and vellum are also untanned skin products, they have physical properties very different to rawhide. These are made by stretching unhaired skins under tension and drying them carefully. As the moisture is removed, they shrink to form an opaque, white, relatively flexible product, which has been employed for millennia as a stable writing or bookbinding material.

Conversely, when the technique of chrome tanning was being developed in the late nineteenth century, it was found that the leather produced would dry out as a hard, stiff, inflexible material. It was only by the addition of fatty lubricating products in the form of an emulsion, in what became known as the fat liquoring process, and by mechanically working the skin that a useable product could be made. In a similar way, the production of alum-tawed skins involved the use of such fatty materials as egg yolk or olive oil and mechanical softening procedures.

It can be seen then that the development of a leathery feel cannot be used on its own as evidence that tannage has taken place.

Shrinkage Temperature.   If a piece of skin, tanned or untanned, is wetted thoroughly, placed in water and heated slowly it will reach a temperature at which it shrinks dramatically to about one-third of its original area. The temperature at which this change occurs is called the shrinkage temperature, and the phenomenon has been likened to melting. It is, though, fundamentally different in that it is irreversible. The shrinkage temperature of any given sample will depend on a large number of factors. These include the species of animal from which the skin was obtained, what pre-tanning and tanning processes it had undergone, the moisture content of the sample and the exact procedures employed for the measurement.

The amount by which a process increases the shrinkage temperature of a skin has been considered as a measure of its leathering ability. Standard methods have been drawn up for the determination of shrinkage temperature and, if they are adhered to, duplicate results within 1° or 2°C can be obtained. Using standard methods, the following shrinkage temperatures are exhibited by typical commercial products:

Raw mammalian skin	58–64°C
Limed, unhaired cattle hide	53–57°C
Parchment	55–60°C
Oil-tanned chamois leather	53–56°C
Alum-tawed skins	55–60°C
Formaldehyde-tanned leather	65–70°C
Aluminium-tanned skins	70–80°C
Vegetable-tanned leather (hydrolysable)	75–80°C
Vegetable-tanned leather (condensed)	80–85°C
Chrome-tanned leather	100–120°C

The majority of these results confirm that while true leathers have higher shrinkage temperatures than raw or pre-tanned skins, materials such as parchment or alum-tawed pelts do not. There is, though, a major exception. Oil-tanned leathers have all the characteristics of true leathers and retain these even after repeated wetting and drying. Their shrinkage temperatures are not, however, increased during the tanning procedures. Oil-tanned leather also exhibits another significant difference in its hydrothermal properties. When other leathers or raw skins shrink in hot water they are converted into a rubbery material, which dries to a hard, brittle product. When oil-tanned skins shrink they retain their leathery feel to a large extent and dry to give a relatively soft and flexible product. Furthermore, if a wet, oil-tanned leather is heated above its shrinkage temperature and then immediately immersed in cold water, it can be stretched back to almost its original size.

The exceptions to the various criteria put forward to define what is, and what is not, a true leather indicate that there is no simple explanation for the tanning effect. It will therefore be necessary to consider the make-up of the skin's leather-making fibres, the nature of their putrefaction and the properties of the wide variety of materials, which are employed to prevent this.

2.2 Collagen

It has been shown that the fibres and fibre bundles which make up most of the skin structure, and which are retained and stabilised during the pretanning and tanning processes, consist mainly of the protein collagen. All proteins are formed from amino acid residues linked together. The properties of any particular protein are determined by which residues are present, and the order and configuration in which they are joined. The collagen in skin exists as long, unbranched, macromolecular chains, each formed from just over 1000 amino acid residues. About 30% of these residues are derived from the smallest amino acid, glycine. A further 10% each are from the five-membered, ring-structured imino acids proline and hydroxyproline. Also present are significant quantities of acidic residues from aspartic and glutamic acids and basic residues derived from, for example, arginine and lysine (Figure 1).

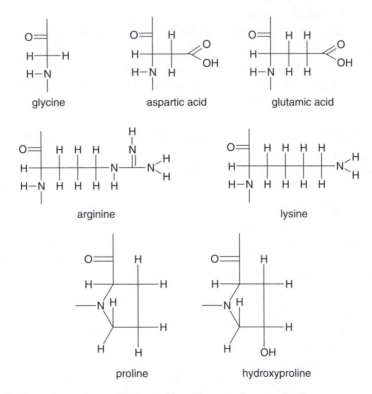

Figure 1 *The major amino and imino acid residue constituents of collagen*

The presence of the ring-structured imino acid residues imparts kinks into the protein chain and the fact that there are so many results in the formation of long, coiled molecules. These have a left-handed twist. The large quantity of small glycine residues enables these coils to be wound tightly. It has been shown that three such helical chains then twist together to give a right-handed helix. The whole structure is stabilised mainly by hydrogen bonding, both within the individual coils, and between the three elements of the helical macromolecule.

The collagen molecule then is a stable, rod-like, triple helix. It is about 300 nm long with a diameter of about 15 nm and has appreciable amounts of potentially reactive free acidic and basic groups on its outer surface. It has been shown that these triple helices then twist together further and cross-link with a quarter stagger to form long, five-stranded, rope-like fibrils with diameters of about 100 nm and these then group and coil to make fibres. A group of fibres are known as fibre bundles.

In addition to the hydrogen bonding which stabilises the triple helix, a variety of links are formed between adjacent molecules within the fibrillar structure. In particular, covalent bonds are formed between the ends of one collagen macromolecule in what is termed the telopeptide region and the helical structure of an adjacent one.

When the animal dies, various mechanisms, which in life protect against the effects of proteolytic bacteria, cease and within hours enzymatic and autolytic reactions commence and the skin begins to putrefy. It has been shown that among the first bonds to be broken are the covalent cross-links binding adjacent molecules together. As a result, the fibrils split apart allowing enzymatic action to cause the triple helices to unwind. This is followed by cleavage along the protein chain itself. As has been mentioned, various methods have been employed to inhibit this bacterial action temporarily. However, to stabilise the skin fully, new cross-links will have to be introduced into the macromolecular structure and these will need to be stable to bacterial attack and not be affected by water. This addition of artificial cross-links is achieved by the tanning process.

It has been generally accepted that hydrothermal shrinkage is a result of the disruption of the hydrogen bonding within and between the polypeptide chains caused by an increase in the molecular vibration of the macromolecule, generated by the introduction of energy to the system in the form of heat. The addition of extra chemically-stable cross-links into the protein complex imparts increased resistance to this molecular vibration, requiring higher levels of energy before these linkages are broken and the structure is disrupted. This is reflected in increased shrinkage temperatures. Recent work on the kinetics of this process has refined this concept and explained for instance why oil tannage produces a true leather without an increase in shrinkage temperature. Tanning, therefore, is the process of introducing additional artificial biochemical-resistant cross-links into the macromolecular protein structure of collagen.

2.3   Tanning Materials

The chemical structure of collagen, a long-chain protein with significant quantities of free acidic, basic and hydroxy-, as well as other reactive, groups on the outer surface of the molecule, allows it to react readily with a diverse range of agents. A number of these are multifunctional, with the ability to form stable links between adjacent protein chains which have the properties required to produce a true leather. Some of the more commercially significant groups will be considered.

Oil and Aldehyde Tannages.   The technique of impregnating skins with oils and fats to make a variety of leather-like materials has been outlined. Among other factors, the properties of the final product are critically dependent on the type of oil used. Brains, marrows and, in particular, marine oils have been found to give a particularly soft, spongy, stable product. It is now known that these contain fatty acids with a medium degree of polyunsaturation. Unsaturated oils and fats such as tallow only lubricate. Highly saturated materials such as linseed oil polymerise around and within the fibre structure, eventually giving a hard, cracky product. Another factor found to be essential for the production of a truly leather-like material is that after the skin is impregnated with the oil, it has to be left in the open air for an extended period of time. It is now known that during this period oxidation takes place at the unsaturated bonds and that this reaction can be enhanced and accelerated by hanging the skins in heated stoves or otherwise treating them with warm air.

The exact nature of the tanning reaction is not known and a number of mechanisms have been proposed. It has been shown, however, that as the process progresses, the degree of unsaturation of the oil reduces, peroxy-derivatives are formed, hydroxyl functions appear and, more specifically, acrolein, $CH_2{=}CH \cdot CHO$, is produced. It is thought that this and other aldehyde compounds are responsible for the chemical cross-linking and that coating the fibres with polymerised oils imparts the special physical characteristics to the leather.

More is known of the mechanism of interaction between the simpler aldehydes, particularly formaldehyde, and collagen where a reaction first takes place at the free amino groups:

$$\text{Collagen—NH}_2 + \text{HCHO} \rightarrow \text{Collagen—NH—CH}_2\text{OH} \tag{1}$$

Crosslinking then occurs between the N-hydroxymethyl group and a free amino group from an adjacent polypeptide:

$$\text{Collagen—NH—CH}_2\text{OH} + \text{H}_2\text{N—Collagen}$$
$$\rightarrow \text{Collagen—NH—CH}_2\text{—HN—Collagen} \tag{2}$$

Formaldehyde was utilised to produce washable white leathers from the end of the nineteenth until the last third of the twentieth century when its use was phased out due to health and safety concerns. Other aldehydes have tanning properties but only glutaraldehyde has been employed successfully on a commercial scale.

Mineral Tannages.   The effects of chromium salts on skins were first investigated during the mid-nineteenth century and by the 1870s, leather was being produced commercially with these materials. Today over 90% of leather is manufactured using trivalent chromium compounds.

The reactions involved in the chrome-tanning process are those of coordination complexes. They involve the interaction between charged carboxyl groups on the collagen macromolecule and polynuclear chromium(III) coordination compounds. The most widely used chrome-tanning material is 33% basic chromium(III) sulfate produced industrially by reducing sodium dichromate with sulfur dioxide.

In aqueous solution, most chromium(III) compounds exist as coordination complexes. Depending on pH, concentration, temperature and other factors these react to form a range of stable basic complexes. Under the conditions usually chosen at the start of chrome tannage (pH 2.5–3.0, temperature 20°C) the predominant species contain two chromium atoms linked with one or more hydroxyl ligands forming ol bridges. Bridging sulfato ligands are also present which further stabilise the structure. At this pH the majority of carboxyl side-groups on the protein chain are not ionised and generally non-reactive. The relatively small chromium compounds can therefore penetrate into the skin structure. As the tannage progresses the pH is gradually raised to 3.5–4.0. This has a dual effect. The protein's carboxyl side-groups become increasingly ionised and the chromium complexes become larger as the number of ol bridges increases. Both these factors enhance reactivity and favour the formation of stable cross-links. These effects are increased by raising the temperature as tannage proceeds, and by the addition of other ligand-forming compounds to control the rate and type of reaction. An outline of some of the reactions, which take place during the tanning process is given in Figure 2.

The use of naturally-occurring potash alum for leather-making dates back at least to pre-dynastic Egypt. Impregnating the skin with pastes of alum, salt, flour, egg yolk or olive oil and water yielded a soft, white product.

As with chromium (III) compounds, aluminium (III) complexes react with collagen's free carboxyl groups. However, under the conditions prevailing during tannage, the complexes do not contain stabilising sulfato ligands and are readily hydrolysed. Nor do the carboxyl groups coordinate so readily into the aluminium complex. The cross-links between adjacent collagen molecules are therefore much weaker and the aluminium salts are easily washed out of

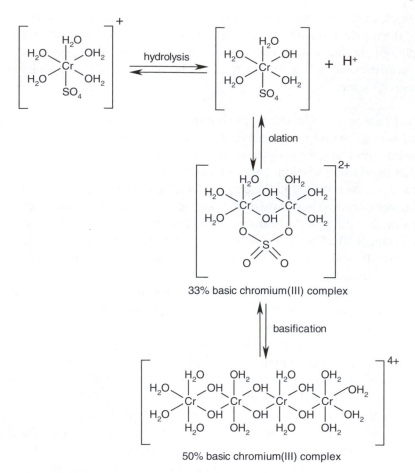

Figure 2 *Formation of polynuclear chromium(III) complexes*

the skins. Thus traditionally alum-tawed materials cannot be classed as true leathers. However, if the aluminium complexes are modified by adding ligands, such as formate or citrate, and the pH is increased, more stable products with true tanning power can be produced. Commercial aluminium tanning materials of this type are available. Aluminium tanned leathers have the advantage of being white but they lack the stability and handling qualities of chrome-tanned products.

Coordination complexes of titanium(III), titanium(IV), zirconium(IV), iron(III) and the lanthanides all have tanning properties, particularly if they are stabilised by the addition of appropriate ligands. None of these, however, give leathers with the properties of chrome-tanned materials.

Vegetable Tannages. Aqueous extracts of barks, leaves, wood, roots and twigs of a large number of vegetable species have been employed to produce

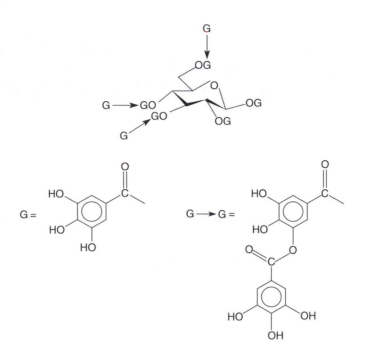

Figure 3 *The major component of Chinese gall extract (tannic acid)*

leather from earliest times. It has been suggested that these were first used to impart colour to oil-tanned or pseudo-leathers. Indeed, until the development of what were known as "chemical tannages" in the nineteenth century, the verb "to tan" was often reserved for the process of vegetable-tanning. It is now known that these aqueous extracts contain many different organic compounds, and that those that react with the collagen are polyphenolic in nature. These are termed tannins and have molecular weights between 500 and 3000. Polyphenols with smaller molecular weights lack tanning power and those with larger are unable to penetrate into the skin's fibre structure.

The vegetable tannins can be divided into two main classes, *i.e.* hydrolysable and condensed. The hydrolysable tannins are obtained commercially from, for example, sumac leaves, tara pods, myrabolam fruits, Turkish or Chinese galls or oak bark. Examples of condensed tannins are those from quebracho wood, mimosa bark and gambier leaves and twigs.

Hydrolysable tannins are more or less complex gallic acid esters of glucose (Figure 3) or compounds related to these based on ellagic acid (Figure 4) or chebulic acid (Figure 5). Their name derives from the fact that in solution they are readily hydrolysed by heat, acids or microbiological action to precipitate out the esterifying acids.

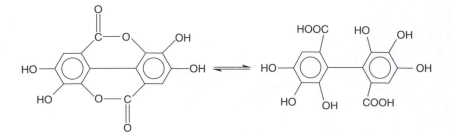

Figure 4 *Ellagic acid*

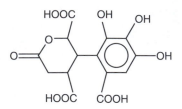

Figure 5 *Chebulic acid*

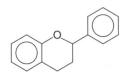

Figure 6 *The flavanoid ring structure on which condensed tannins are based*

The condensed tannins consist of flavinoid groups (Figure 6) containing various combinations of phenolic hydroxyl groups linked together to form di-, tri- or larger polymers (Figure 7). Unlike the hydrolysable tannins, which are broken down by hydrolysis, the condensed tannins polymerise further to form high molecular weight, insoluble complexes.

The reaction between vegetable tanning materials and collagen is mainly due to the large number of phenolic hydroxyl groups on the tannin molecules. These form hydrogen bonds with the electronegative centres on the protein chain, such as the carboxy- and imino groups. There is also the possibility of reaction between the polyphenols and the amino or carboxylic side-chains depending on pH.

Syntans. With the development of an understanding of the polyphenolic nature of vegetable tannins, and the increasing shortage of these naturally-occurring materials at the beginning of the twentieth century, efforts were made to synthesise products with related chemical structures in the hope that

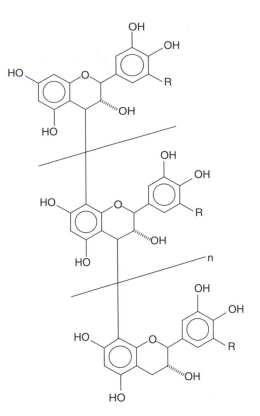

Figure 7 *Linking of flavanoid-based polyphenols to form condensed tannins*

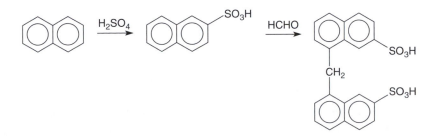

Figure 8 *Outline of the Neradol method for syntan production*

they would have similar tanning properties. The first commercially success-ful products were produced by sulfonating aromatic compounds containing hydroxyl groups and then polymerising them with formaldehyde in what has become known as the Nerodol process (Figure 8).

Later the Novolac process was developed in which the phenolic compound was first polymerised and the product sulfonated to the required level (Figure 9).

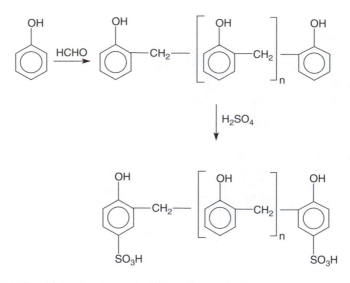

Figure 9 *Outline of the Novolac method for syntan production*

With the replacement of formaldehyde with such polymerising agents as urea, dimethyl methylene and ether and the introduction of a more complex variety of aromatic base groups, a very wide range of synthetic tanning materials (syntans) have become available commercially.

The tanning reaction of syntans is partially due to the presence of hydroxyl groups forming hydrogen bonds in a similar manner to vegetable tannage, but is mainly a result of the sulfonate groups forming electrovalent bonds with the free amino groups on the collagen.

3 THE MANUFACTURE OF LEATHER

The first leathers were probably made by our hominid ancestors using oils and fats, possibly in conjunction with a smoking stage. It has been speculated that scraping the skin impregnated the structure with subcutaneous adipose material giving a softer, more long-lasting product. This led to other fatty products from the animal, such as brains and marrows, being used. These contain both unsaturated fatty acids and natural emulsifying agents resulting in the production of a true leather. It has also been speculated that it may have been observed that skins used to cover tents in which smoking fires were lit gradually developed further desirable properties and that this resulted in the introduction of a separate smoking process.

Common salt (sodium chloride) has been used to preserve food since earliest times. It is, therefore, quite possible that naturally-occurring alum was first employed to treat skins when it was mistaken for common salt. The special properties of these particular crystals were noted and exploited for making leather, as well as for other purposes, long before the pre-dynastic Egyptian period.

It has been suggested that vegetable tanning developed from a desire to colour oil or alum-processed skins. Interestingly the earliest surviving recipes for the preparation of leather, dating from Babylonian times about 3000 years ago, show that a combination of these three processes were employed:

> You will steep the skin of a young goat with the milk of a yellow goat and with flour. You will anoint it with pure oil, ordinary oil and the fat of a pure cow. You will soak the alum of the land of the Hittites in grape juice and then cover the surface of the skin with gall nuts collected by the tree growers of the land of the Hittites.

Certainly by the classical Greco-Roman period, vegetable tannage had developed into an important craft-based industry and little fundamental change appears to have occurred until the beginning of the nineteenth century.

The great majority of leather artefacts surviving in museums, historic houses and private collections will have been vegetable tanned. This, therefore, is the predominant material that conservators will be required to deal with. The method of manufacture of this product will therefore be considered first. Significant changes were made during the late nineteenth and early twentieth centuries in the processes employed to make vegetable-tanned leathers. Some of these have had serious consequences regarding the rate of deterioration and it is this class of leather which most frequently requires conservation treatment today. These changes will therefore be discussed separately.

3.1   Vegetable Tanning Processes

Medieval and Pre-Modern Period.   The first job of the tanner was to wash the hides free from blood and dung. This was often done by immersion in the local stream. It was then necessary to loosen the hair to enable it to be scraped off without damaging the grain surface. The most primitive method was to pile the hides until putrefaction set in just enough to loosen the hair roots. An alternative method was to soak the skins in alkaline liquors prepared from wood ash or, more commonly, lime. A third method can be considered as a combination of the first two. In this, the hides were immersed in lime liquors which had been "mellowed" by repeated use. These liquors would have contained large concentrations of organic breakdown products which would have accelerated the depilatory action. The hide was then spread over a wooden beam and both sides scraped with tanners' two-handled knives. The hair side was scraped with a blunt, unhairing knife and the flesh with a sharper, two-edged, fleshing knife.

The pelts were then given a further cleansing and opening up treatment using the alkaline bating, puering or mastering process or the acidic raising or drenching process. In the alkaline processes, the pelts were immersed in a warm infusion of bird droppings or dog dung. This removed excess lime and biochemically altered the hide structure to give a softer leather with a finer, more flexible grain. In the first decade of the twentieth century the active ingredients in the dung were found to be proteolytic enzymes secreted in the pancreas and activated by ammonium salts. The acidic raising or drenching process involved treating the hides in liquors prepared by fermenting barley, rye and other vegetable matter. The action of the fermentation was to produce a complex mixture of organic acids and enzymes which broke down the non-fibrous proteins of the skin and removed excess lime. Once the pelts were judged to be in the correct condition, the hides were again washed and worked over the beam with a blunt knife to remove the slime that had been liberated. These pre-tanning processes could have taken up to three months.

The preliminary stage of the tanning operation itself was to immerse the hides in weak, almost spent, tanning liquors, moving them around continuously. Once the colour of the grain was judged to be satisfactory, the hides were transferred to a further set of pits. To prepare these, a layer of ground vegetable-tanning material was tipped into the bottom of the pit and a hide laid flat over it. The vegetable-tanning material employed depended on what was available locally. Birch, willow, spruce and larch were used in Northern Europe and Russia; various species of oak in Britain and Central Europe; sumac, valonia, oak galls and various acacias around the Eastern Mediterranean. A layer of ground tanning material was strewn over the hide, then a second hide spread out over it. In this manner alternate layers of tanning material and hides were added until the pit was nearly full. A final layer of tanning material was piled on top and the whole pit filled either with water, or an infusion prepared by extracting tanning materials with cold water. The hides were generally kept in these pits for at least a year, and the whole layering process could be repeated up to three times. When the tanner had judged the hides to be fully tanned, they were rinsed off and smoothed out. The leather was then dried out slowly in a dark shed, fitted with louvered panels to control the rate of drying.

The dried rough leather was sold to a currier whose first operation was to dampen the skin and soften it. This was followed by a scouring operation in which the leather surfaces were scrubbed clean. The skins were then pared down to the required thickness after which they were worked on a bench to remove loose tanning materials, flatten and stretch them. They were then partially dried and impregnated with a warm mixture of tallow and fish oils. If a soft, fine product was required, further mechanical operations were undertaken. These had the effect of separating the tanned fibres and allowing them to move over each other. If required, the leather was coloured, using natural dyestuffs, often in conjunction with an alum mordant. It was only after all these operations had been carried out that the leather was ready to be sold.

Nineteenth Century. While the changes that took place in processing during the 19th century were considered to be improvements on previous techniques aimed at giving a less expensive, more uniform product, at the same time some had deleterious effects which only became apparent decades after they were introduced. This was particularly significant in the case of leathers for bookbinding which, unlike those for other purposes, are expected to last a lifetime. This is also relevant to leather objects which are stored and displayed in museums and other collections, long after the time when they would have normally been expected to have been discarded. These potentially damaging changes in tanning techniques will be considered.

The rapid increase in population and improvements in standards of living which took place in the nineteenth century led to a rise in demand for leathers of all types, resulting in shortages of both indigenous tanning materials and hides. This encouraged the introduction of a wider range of imported tanning materials, including many species not previously employed in Europe. We now know that leathers prepared using condensed tannins decay more rapidly than those manufactured using hydrolysable tannins. It is fortuitous therefore that the new imported materials which gave the best quality light leathers, were mostly of the hydrolysable type. It is quite possible, however, that condensed tanning materials were employed by some tanners, resulting in leathers with poor ageing properties.

A potentially more damaging improvement was the introduction of synthetic dyestuffs from the middle of the century onwards. Initially, these were developed for colouring textiles, but soon they were being used on a wide range of leathers. The replacement of the traditional natural materials with these new products adversely affected the ageing properties of the leathers in two ways. First, the natural dyestuffs previously employed were almost invariably applied in conjunction with an alum mordant whose beneficial effects are now well known. The new dyestuffs were either used alone or together with mordants, such as potassium bichromate, iron sulfate or tin chloride which, unlike alum, actually have a damaging effect. Second, it was found that in order to "fix" the dyestuff to the leather and "clear" the dyebath it was necessary to increase its acidity. Sulfuric acid was generally used with extremely deleterious results. It is now known that sulfuric acid either derived from atmospheric pollution or added during processing has been the major cause of the deterioration of bookbinding leathers and that the presence of strong acids, whatever the source, is particularly damaging.

Just as the demand for leather outstripped the supply of vegetable-tanning materials, there was an increasingly serious shortage of skins for the tanners to process. This problem was overcome by the importation of crust leathers which had been roughly tanned in their country of origin either using local, traditional methods or local variations of "European" systems. Indigenous tanning materials were employed, many of them of the condensed type. These skins were usually tanned locally on a very small scale and gathered together from over a wide area to a centralised trading depot. Here they were sorted, mainly into size and quality, and exported. Once in Europe, they were often sorted again before selling on to a tanner or leather dresser. Crust skins were sold by weight and often contained adulterants, such as excess tannins, oils, earthy materials and soluble salts. In addition, the dresser was purchasing a very mixed lot of skins from a wide range of sources having very different processing histories. It was his job to minimise these differences and produce as uniform a batch of leather as possible.

In order to achieve this, he first washed the skins thoroughly to remove as much of the unwanted weighting material as possible. He often added alkalis to the wash water in what is known as the stripping process. This was designed to remove excess tannins and to unify the dyeing properties of the mixed batch of leather but, unfortunately, it also removed non-tans. These include various organic salts which are present in the leather but do not actually contribute to the tanning reaction. They are, however, effective buffers and have a protective action against the action of acids.

Next, the skins were shaved to the required thickness. By the last quarter of the nineteenth century various shaving machines had been developed which replaced the laborious, skilled, hand operations. With these machines it was possible to cut skins down to the required thickness, cheaply and accurately. One feature of these machines, though, was that in order to ensure a clean cut, their blades had to be sharpened continuously. This resulted in a shower of sparks and small specks of iron dropping onto the leather. These caused a pattern of blue-black iron stains, which had to be removed in the clearing process. In this, the skin was immersed in a solution of sulfuric acid, which not only dissolved the iron salts but changed the colour of the leather from a reddish brown to a pale yellowish buff. Once again, dangerous sulfuric acid was introduced into the leather. In order to produce a more uniform substrate for the dyeing process, the leathers were then retanned, possibly using tanning materials of the condensed type. They were then dyed, often with synthetic colours, acidified, again with sulfuric acid, dried out and finished.

By the last decade of the century, the effects of these changes had become only too apparent, and efforts were made to determine the causes of the problems and eliminate them. Conservators are, however, still working today on objects made from leathers manufactured during the period when the long-term effects of these "improvements" had not been appreciated.

3.2 Alum Tawing

Until the beginning of the nineteenth century vegetable-tanned leather was manufactured by the tanner and the currier who generally worked on larger cattle hides. Sheep, goat, deer and dog skins were generally processed by the whittawyer or the glover. Fur-bearing pelts were prepared in a similar manner by the skinners.

The pre-tanning processes employed by the whittawyer were similar to those of the tanner. Once the skins had been limed, unhaired, fleshed and given a thorough bating they were put into large wooden tubs. There they were kneaded with a mixture of alum and salt, a lubricant such as egg yolk, butter, or olive oil and a carrier such as flour or oatmeal. Traditionally the tawyer worked the mixture into the skins by trampling it in with his bare feet. Once the required amount of the tawing paste had been taken up by the skins, they were stretched out flat and piled overnight. The next day they were mechanically worked again and hung to dry. The leather was then softened by working them mechanically.

3.3 Oil Tannage

Widely-used alternatives to the alum-based tawing pastes were the various oxidisable marine oils, which were applied in the production of the chamois and buff leathers. The oils were trampled into the skins in a similar manner to that used in tawing. The skins were then hung in warm, airy stoves for the oxidation process to take place. After the oiling and stoving sequence had been repeated three or four times, the leather was washed off in alkaline liquors to remove excess oil, dried and worked mechanically to soften them.

3.4 Fur Dressing

Skinners processed pelts from a variety of fur-bearing animals ranging from rabbits and cats to wolves and bears. The skins were usually obtained in a dried condition and required careful wetting back before excess flesh could be removed and the pre-tanning operations undertaken. Great skill was required to ensure that the structure of the pelt was opened up sufficiently, while at the same time the hair remained firmly attached. Skinners employed the alum- and oil-tanning processes similar to those used by whittawyers. Alternatively, fur skins were treated in baths of fermenting grains similar to those described in the raising or drenching processes used by tanners and dried without further processing. The organic acids and enzymes helped to remove the non-collagenous interfibrillary materials and give a thinner, softer product. These drenches did not tan the pelts but made them acidic enough to prevent bacterial attack.

4   THE DETERIORATION OF LEATHER

Leather and other skin products, in common with all other materials, organic or inorganic, are subject to change and decay. The rate of this change is dependent on the product itself (raw material, type of tannage, *etc.*) and the environment to which it is subjected. Deterioration may be caused by physical, biological or chemical agents or a combination of these. This section will highlight some of the major factors leading to decay in leather and, where they are known, will discuss the mechanisms involved.

4.1   Physical Deterioration

Leather objects are likely to show evidence of the wear and tear that they were subjected to before they entered the museum environment. This will include splits, tears, scratches and holes, as well as damage resulting from the effects of perspiration, urine and accidental spillage of other undesirable liquids. This type of mechanical deterioration does not automatically cease when the object is placed in a collection, as damage caused by inappropriate handling is not unknown.

Leather and other skin products have the ability to absorb and desorb large amounts of water vapour from the surrounding atmosphere without appearing wet. This property is associated with the amount of hydrogen and water bonding within the triple-helical collagen structure. As leather takes up moisture from a humid atmosphere, it increases both its thickness and, particularly, its area. In some cases these can increase by over 7% if they are moved from an environment with a relative humidity of 20% to one of 80%. If a piece of leather has been placed under restraint, for example, by fixing it to a wooden frame, and it is moved to a low humidity environment, it will attempt to shrink. This will set up tensions that could result in splitting and tearing.

A related phenomenon is termed age hardening. This, as the name suggests, exhibits itself by a progressive stiffening and shrinkage of the leather. In more severe cases the stresses caused by differential shrinkage of the outer and inner layers of the skin result in cracking and, eventually, loss of the whole grain layer. The main cause of this condition is thought to be excessively high, or excessively low, moisture content in the leather and, in particular, cycling between these two conditions.

The mechanism of deterioration appears to be that, during periods of high humidity, the excess moisture present in the leather dissolves soluble materials. As the leather dries out, the moisture moves to the surface, transporting these materials with it, leaving them as a deposit on the surface structure as it evaporates. Over time, the removal of the tannins leads to a detanning of the inner layers as is evidenced by a drop in shrinkage temperature. This causes shrinkage and stiffening. The presence of excess material in the grain structure reduces flexibility and increases the tendency to grain crack.

4.2 Biological Deterioration

By definition, true leathers are resistant to bacterial attack. If, however, rawhide, parchment or pseudo-leathers become wet, putrefaction sets in and they will decay rapidly due to the action of proteolytic enzymes secreted by the bacteria. Even prolonged exposure in very damp atmospheres can render parchment open to bacterial attack.

While leathers are resistant to bacteria, the same cannot be said about fungi. More than 50 species of fungi have been reported to have been found growing on leather, and much work has been carried out by the leather industry to determine the factors favouring mould growth. It has been reported that: leathers tanned with vegetable extracts are much more liable to mould growth than chrome-tanned materials; vegetable-tanned leathers processed with hydrolysable tannins are more susceptible to attack by fungi than those tanned with condensed tannins; and leathers with a high fat content are more open to fungal attack than those containing only small amounts of fatty materials.

It has also been determined that the loss of strength, tendency to crack and other signs of physical deterioration associated with mould growth are not due to proteolytic attack. The fungal enzymes tend to digest fats, sugars and other carbohydrates present in the leather, liberating various organic acids. It is these acids that react with the vegetable tannins and break down the tannin–protein complex.

Common names such as the hide beetle, the bookworm and the leather beetle suggest that leather is particularly susceptible to attack by insects. This, however, is not the case. Untanned collagen is a rich source of protein for a large number of insects but, once it is tanned, collagen is not broken down by most of the proteolytic enzymes secreted in the insects' digestive systems. Insect damage to leather objects is, therefore, usually coincidental as the insect passes through the leather structure or along the surface in search of something more nourishing. The bookworm or booklice, for instance, feed on cellulose-based products such as paper or, in particular, the starch-based adhesives used to attach leather bindings to book boards. The various wood-boring beetles are also quite capable of passing through leather on their way in and out of a wooden structure underneath.

From time to time, conservators are asked to repair physical damage caused by vertebrate pests. In the author's experience these include mice, squirrels, cats, dogs, small boys and a parrot. No doubt other conservators can add to this list.

4.3   Chemical Deterioration

Leather can be considered as a relatively chemically-stable material. In general objects made from leather reach the end of their useful lives as a result of fair wear and tear rather than chemically-induced decay. If a pair of shoes lasts for 10 years or a wallet or briefcase for 20 years, the owner is usually satisfied. Books are different. Many leather-bound books spend the majority of their lives sitting unused on shelves. Libraries in historic houses, for instance, contain thousands of books which are rarely read but whose bindings remain sound after a century or more. It is not surprising, therefore, that when in the middle of the nineteenth century it was discovered that many new bindings were deteriorating rapidly, an explanation was sought.

The first systematic investigation into the problem was undertaken for the Athenaeum Club in 1841 by a committee of tanners and chemists chaired by Michael Faraday. The cause of the rapid decay was shown to be sulfur dioxide given off by the newly installed coal gas reading lamps, which oxidised and hydrated to form sulfuric acid within the leather fibrous structure. As the century progressed leather bookbindings from libraries without any gas lighting were also found to suffer from the same form of deterioration, which became known as red rot. Further work was carried out under the auspices of the Royal Society of Arts. Their report confirmed that the major cause of red rot was the presence of sulfuric acid in the leather. This could either have come from sulfur dioxide pollution in the atmosphere or have been added during the tanning processes. They also concluded that leathers tanned with hydrolysable tannins were superior in their ageing properties to those processed with condensed tannins.

In the 1930s, a major study was undertaken to determine the effects of different environmental conditions on over a hundred commercially- and experimentally-prepared leathers. Duplicate sets of books were bound with these. One set was shelved in the British Library, representative of conditions in a polluted urban atmosphere and the others in the National Library of Wales in Aberystwyth, a clean rural setting. The books were examined at regular intervals over a period of nearly 50 years for evidence of degradation.

The results showed that while none of the books stored in the clean Welsh environment had deteriorated to any significant extent, nearly all the volumes subjected to London's acidic pollution exhibited evidence of decay, some within less than 10 years. In addition, the superior ageing properties of leathers prepared with hydrolysable tannins were confirmed. Of the various chemical analytical determinations undertaken, the only results to show any correlation with the degree of deterioration observed were those for the number of *N*-terminal amino acid groups on the protein. This figure reflects the amount by which the collagen polypeptide chain had been broken and was considered to be evidence for hydrolytic deterioration.

Further work over a long period confirmed these general conclusions. It also showed that retanning vegetable-tanned leather with mineral-tanning materials, particularly aluminium salts, gave increased resistance to acid deterioration and this led to the formulation of a British Standard for Archival Quality Bookbinding Leathers. Unfortunately, imparting the desired chemical resistance to acidic deterioration was associated with a loss of the specific physical handling characteristics required by bookbinders.

From the beginning of the twentieth century it has been suggested that chemical deterioration of leather is due to a combination of oxidation and hydrolysis. Research undertaken by a group of conservation scientists funded under the European Commission STEP and ENVIRONMENT programmes has confirmed and refined this hypothesis. These projects examined the changes that occur to the amino acids and tannin polyphenols as vegetable-tanned leathers are aged naturally and artificially. From the many results published the following appear to be the main conclusions.

In less polluted environments, deterioration is associated mainly with oxidative change. This can be activated by the effects of heat, light or the action of free radicals. The latter can arise from the autoxidation of lipids or the breakdown of tannin polyphenols. The main effect on the collagen macromolecule is to convert the basic side groups of lysine, hydroxylysine and arginine to acidic groups, such as glutamic and α-amino adipic acids (Figure 10). In a similar manner, proline and hyroxyproline can be oxidatively degraded to give glutamic and aspartic acids (Figure 11). Leathers naturally aged under relatively unpolluted conditions were therefore found to contain reduced quantities of basic amino acids and higher amounts of acidic ones compared with new leather. They were also found to contain α-amino adipic acid, which was absent in new leather. In heavily polluted atmospheres, on the other hand, the main effect appears to be the direct acid hydrolysis of the bonds between amino acid residues in the main protein chain (Figure 12).

Interestingly, it was found that sulfur dioxide was absorbed twice as rapidly from polluted atmospheres by leather produced with condensed tanning materials than those processed with hydrolysable tannins. It was also noted that

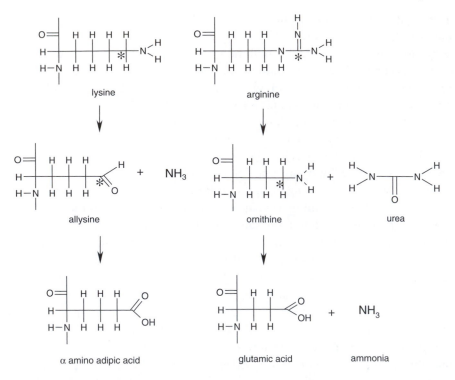

Figure 10 *Possible oxidative mechanisms for the production of acidic side groups from basic amino acid residues. Asterisk * indicates site of oxidative attack*

there was evidence that in highly polluted atmospheres the acidic environment suppresses oxidative reactions.

It can be concluded that chemical degradation of vegetable-tanned leather will occur in both polluted and unpolluted environments, and that both oxidative and hydrolytic reactions are always involved. The preponderant mechanism will depend on the exact conditions to which the leather is subjected. It should also be noted that the rate of decay of leathers exposed to the mainly hydrolytic effects of acidic urban atmospheres is much greater than the oxidative deterioration found with leathers held in cleaner rural surroundings.

Research has been carried out to determine the mechanism of deterioration of parchments. This has shown a very similar combination of oxidative and hydrolytic attack.

Further investigations, funded by the European Commission CRAFT programme, have also been undertaken to develop a leather with both the resistance to acidic atmospheric attack exhibited by mineral tanned leathers and the physical characteristics of vegetable-tanned skins demanded by book binders and other leather workers. As a result of this work, leathers with these properties are now available commercially.

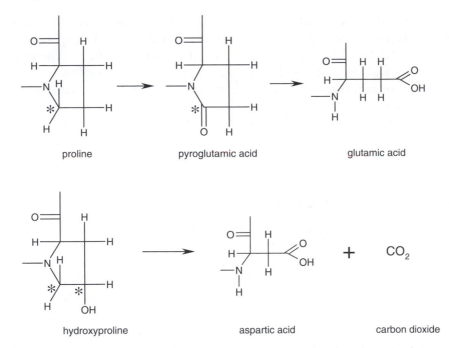

proline → pyroglutamic acid → glutamic acid

hydroxyproline → aspartic acid + carbon dioxide

Figure 11 *Possible oxidative mechanisms for the production of acidic side-groups from imino acid residues. Asterisk * indicates site of oxidative attack*

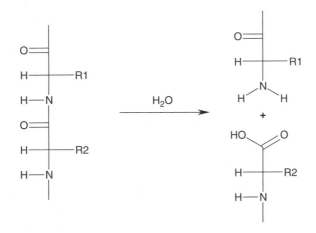

Figure 12 *Hydrolytic scission of protein chain*

5 CONSERVATION TECHNIQUES

Conservation can be considered to be the application of a treatment or a series of treatments to an object of artistic, historic or even sentimental value to prevent, halt or reverse deterioration. The more philosophical and ethical considerations of the profession have been discussed, among others, by Pye and the practical application of these considerations to leather objects by Sturge.

The conservation treatments applied to a leather object may include recording, pest eradication, cleaning, reshaping, repairing, reinforcing, infilling, inpainting, consolidation, chemical stabilisation and surface coating. The choice of techniques and materials employed for each of these stages is often based on previous, scientifically-based investigative work. This includes studies such as those into the strength and flexibility of repair materials, the stability of adhesives and the penetration of consolidants into fragile substrates. This section will, however, concentrate on treatments developed for the major causes of deterioration of leather objects.

5.1   Biological Deterioration

As is the case with most conservation strategies, preventing biological attack is far preferable to treating the results. It is known that at temperatures normally experienced in temperate climates, such as Northern Europe, mould will not grow on leather at relative humidities of less than 65%. Leather objects should therefore be kept away from damp walls, and efforts should be made to circulate the air in display and storage spaces to prevent the build up of moist, damp areas. In many countries, relative humidities in excess of 65% are commonly experienced and the cost of dehumidification prohibitive. In these cases, the only option is the use of a suitable fungicide. A large number of these have now been withdrawn because of health and safety regulations. The products currently most widely employed in the leather-producing industry to prevent mould growth are based on 2(thio cyano methyl thio) benzothiazole, TCMTB. Despite the somewhat chemophobic attitude of many curators and conservators, this material has been used successfully where climate control is impossible.

Just as the routine use of chemical fungicides is being avoided, insecticides are now only being employed where absolutely necessary. The preferred techniques are the use of cold, heat or anoxic atmospheres. Freezing the object to −25°C for a period of 24 h has been shown to kill virtually all insects. Care must be taken to wrap the object properly to prevent the formation of potentially damaging condensation as it warms back to ambient temperatures. Heating an object approximately 50–55°C will also kill all insect pests. Provided the humidity of the object and its surroundings is carefully controlled as with the Thermo Lignum process, a wide variety of skin materials can be satisfactorily treated in this way.

5.2   Chemical Deterioration

As has been stated, the most widespread form of chemical deterioration found in historic leather collections is the result of a combination of oxidative and hydrolytic degradation of vegetable-tanned skins, leading to what is known as red rot. This form of decay is associated with the presence of strong acids, particularly sulfuric acid, in the leather. It has been found that a vegetable-tanned leather with a pH of over 3.2 is unlikely to be suffering actively from this form of deterioration. Those with a pH of 2.8 or less are either exhibiting the symptoms of acid deterioration or will probably develop them in the near future. This finding led to a number of modifications being incorporated into the manufacturing process employed to make furniture, bookbinding and other leathers which might be expected to have an extended life. These were aimed at neutralising any acids present or making the leather resistant to acidic atmospheric pollutants. The use of strong alkalis had to be avoided as increasing the pH of a vegetable-tanned leather above 5.5 can cause darkening, detannage and embrittlement. Salts such as lactates, citrates, phthalates and tartrates, which buffer leather between 4.0 and 5.0, were therefore employed.

A number of techniques using similar concepts were developed over the years to treat leathers that had already deteriorated. Aqueous solutions of potassium lactate were routinely applied to the surface of bookbindings and other leather objects. Unfortunately, in many cases, the deleterious effects of the liquid water solvent on acidic leathers took place before the buffer salt neutralised the acid, resulting in dark, seriously embrittled leather. In order to avoid the effects of aqueous solutions, methods were tried employing ammonia vapour. The object was placed in an enclosed chamber over an open dish of ammonium hydroxide solution. However, ammonia vapour is a strong alkali and there is a danger that even with the use of dilute solutions, leathers that were too acidic would be transformed into ones which were too alkaline. Another alternative was to use an organic base such as imidazole in a non-aqueous solvent. However, this was also liable to result in leathers that were too alkaline.

It has been shown that by retanning vegetable-tanned skins with aluminium salts, leathers could be produced which resisted the effects of acidic atmospheric pollution. It was thought, therefore, that treating acid-deteriorated leathers with solutions of aluminium compounds such as sulfate, chloride or formate could have a beneficial effect. Trials using artificially-aged leathers showed that the acid was neutralised and the pH was raised. Moreover, the shrinkage temperature of the damaged leather was increased showing that the aluminium salts were reacting with the partially-degraded collagen, introducing new chemical stabilising bonds. However, when the system was applied to naturally damaged leather, it was found that, as with the use of aqueous buffer solutions, in a significant proportion of cases the damaging effects of water occurred before the beneficial protective reactions of the aluminium compounds could take place.

As a method of avoiding the damage caused by aqueous solutions, an attempt was made to find an aluminium compound which was both soluble in a non-polar solvent and had significant tanning powers. A wide range of compounds was screened and one, an aluminium isopropoxyde derivative chelated with ethyl acetoacetate called, for simplicity, aluminium alkoxide, was found to have the optimum combination of properties. This is used as follows.

A solution of aluminium alkoxide in white spirits is applied to the acid damaged leather and the solvent is allowed to evaporate. The reaction takes place over an extended period of time depending on temperature and humidity and is thought to proceed thus: the organic complex does not react with the leather and can therefore penetrate fully into the fibre structure; the solvent then evaporates and water, in the form of moisture present in the leather, reacts with the compound, changing it from an organic to inorganic co-ordination complex. It is this complex which reacts with both the degraded vegetable tannins and the partially denatured proteins, forming new chemical links between the tannins and the collagen, in effect retanning the leather. This is indicated by an increase in the shrinkage temperature. The inorganic aluminium compound also reacts with free acid, buffering the leather at about pH 4.5.

All conservation treatments should ideally be reversible. The use of aluminium alkoxide is most definitely not. It should be noted also that while treatment with aluminium alkoxide improves chemical stability, there is no improvement in physical characteristics and the use of suitable consolidation treatments should be considered.

6   CONCLUSIONS

Leather can be defined as a material made from the outer skin of any vertebrate (fish, reptile, bird, mammal, *etc.*) by any process which renders it non-putrescible under warm, moist conditions. A true leather will retain this fundamental property even after repeated wetting and drying. These properties are imparted to the main constituent of the skin, the protein collagen, by introducing additional microbiologically-stable cross-links into the polypeptide structure.

It will be seen that the successful treatment of any object requires the conservator to have a wide range of skills. These include an understanding of the scientific principles behind any treatment process, a knowledge of the technology of the material from which the object was made, an appreciation of the techniques employed in its manufacture and the craft skills to undertake the conservation procedure. This applies just as much to the treatment of leather artefacts as to those which are sometimes more conventionally valued.

REFERENCES AND FURTHER READING

General

A.D. Covington, *Chem. Soc. Rev.*, 1997, **26**, 111.
M. Kite and R. Thomson, *Conservation of Leather and Related Materials*, Elsevier, Oxford, 2006.

Properties of Leather

A.D. Covington, *J. Soc. Leather Technologists and Chemists*, 2001, **85,** 24.
B.M. Haines, *The Fibre Structure of Leather*, The Leather Conservation Centre, Northampton, 1981.
R.M. Lollar, in *Chemistry and Technology of Leather*, vol. 2, F. O'Flaherty, W. T. Roddy and R. M. Lollar (eds), Reinhold, New York, 1958, 1.
Y. Nayudamma, in *Chemistry and Technology of Leather*, vol. 2, F. O'Flaherty, W. T. Roddy and R. M. Lollar (eds), Reinhold, New York, 1958, 28.

Collagen

C.J. Kennady and T.J. Weiss, *Restorator*, 2003, **24**, 61.

Tanning Materials

J.H. Sharphouse, *J. Soc. Leather Technologists and Chemists*, 1985, **69**, 29.
T.C. Thorstensen, in *Chemistry and Technology of Leather*, vol. 2, F. O'Flaherty, W. T. Roddy and R. M. Lollar (eds), Reinhold, New York, 1958, 221.
T. White, in *Chemistry and Technology of Leather*, vol. 2, F. O'Flaherty, W. T. Roddy and R. M. Lollar (eds), Reinhold, New York, 1958, 98.

Manufacture of Leather

E. Cameron (ed), *Leather and Fur: Aspects of Early Medieval Trade and Technology* Archetype, London, 1998.
R. Reed, *Ancient Skins, Parchments, and Leather*, Seminar Press, London, 1972.
R. Thomson, *Trans. Newcomen. Soc.*, 1981–1982, **53**, 139.
R. Thomson, *J. Soc. Leather Technologists and Chemists*, 1991, **75**, 85.
E.M. Veale, *The English Fur Trade in the Later Middle Ages*, Oxford University Press, Oxford, 1966.

Deterioration

R. Larsen (ed), *STEP Leather Project*, Royal Danish Academy of Fine Arts, Copenhagen, 1994.

R. Larsen (ed), *ENVIRONMENT Leather Project*, Royal Danish Academy of Fine Arts, Copenhagen, 1997.

R. Larsen (ed), *Microanalysis of Parchment*, Archetype, London, 2003.

R. Thomson, in *Pest Attack and Pest Control*, United Kingdom Institute for Conservation, London, 1996, 34.

R. Thomson, *J. Soc. Leather Technologists and Chemists*, 2001, **85**, 66.

R. Thomson, *Leather Conservation News*, 2003, **17**, 1.

Conservation

E. Pye, *Caring for the Past: Issues in Conservation for Archaeology and Museums*, James and James, London, 2001.

T. Sturge, *The Conservation of Leather Artefacts*, The Leather Conservation Centre, Northampton, 2000.

R. Thomson, *Leather Conservation News*, 1995, **11**, 11.

CHAPTER 6

Metals

DESMOND BARKER

Consultant, Formerly Department of Applied Chemistry,
University of Portsmouth, UK

1 INTRODUCTION

Some of the earliest known metallic artefacts probably originated from nuggets
of pure metals such as copper or meteoric iron. A dagger dating from 3000 BC
is a good example of the latter, which gave a chemical analysis of 90% iron
and 10% nickel. Metals are rarely found in their native form and exist in the
earth's crust as minerals predominately as sulfides or oxides. Man soon discovered that it was possible to produce molten metal in containers (crucibles) by
making their fires hot enough. This molten metal could then be poured into a
shaped cavity made by placing together two halves of hollowed out clay or
stone mould. The metal took the shape of this cavity when solid. This process is
called "casting". A bell recovered from 1000 BC is an early example of an
artefact made in this manner. It was a mixture of copper and 13% tin that is
known as bronze. Countless examples of this alloy have been excavated,
although the tin content does vary considerably. Early metallurgists soon
found that by changing the alloy content, the mechanical properties could be
altered. Lowering the tin content, the bronze became softer, while increasing it
the alloy became harder.

   By the time that the Romans came to Britain, iron and bronze were being
used for weapons and tools; copper for vessels and ornaments; lead for pipes
and coffins; silver and bronze for coinage; and gold, silver and tin for ornaments. A considerable number of Roman drinking vessels have been found
made of pewter, which is an alloy of lead and tin. Even the Romans must have
known something about lead poisoning as their superior quality pewter had a
lower level of lead in the alloy; 20% instead of the normal 50%.

   Brass is an alloy of copper and zinc and although Roman coins have been found manufactured from this alloy (71% Cu, 28% Zn), it was not until the fourteenth century that reasonable quantities of this alloy could be made.

   Iron and its alloys (ferrous metals) were slow to develop because of the difficulty in winning the metal from its ores. Most of the early metals and alloys had relatively low-melting points (e.g. lead 327°C), while iron has a melting point of 1535°C. This temperature was impossible to achieve until the nineteenth century. Hence, only solid masses (blooms) were produced that were subsequently hammered into shape – forming wrought iron. By the sixteenth century, it was observed that the formation of an iron/carbon alloy (cast iron) lowered the melting point to approximately 1200°C. This allowed the production of cast iron guns, which were considerably cheaper than their bronze counterparts. Cast iron was very brittle and cracked fairly easily if subjected to impact loads. This was due to the high carbon content (~4% carbon) in the alloy, which was predominately in the form of graphite flakes. It was not until the middle of the nineteenth century that large quantities of iron with less than 1.7% carbon could be melted to produce an alloy called steel. The result of this advancement in technology signified the beginning of the industrial revolution and the commencement of the wide-scale use of metals and alloys.

   The minerals from which the metals are extracted, existed for millions of years in the earth's crust and are the most stable form of the metal. A considerable amount of energy is required to convert this mineral into the metal. Once this pure metal comes into contact with the natural environment such as sea-water or soils, the metal slowly converts back to its original starting material. Iron, for example, is obtained from the mineral, haematite, an oxide of iron. Once the pure iron comes into contact with water and air (oxygen), it slowly converts back to the oxide. This is called corrosion and the product is familiar to everyone as red rust. Nearly all metals will corrode in natural environments although the rates of corrosion will vary from metal to metal and alloy to alloy. In addition, the rates of corrosion will vary from one natural environment to another. Iron will corrode at approximately 50 μm per year in freshwater but at 120 μm per year in seawater. The reason for this is due to the difference in chemical composition between freshwater and seawater. The latter contains salt (sodium chloride) and this is very deleterious to the corrosion behaviour of the metal. Silver artefacts may be excavated after several hundred years buried in soils with only minimal amounts of corrosion. Those recovered from marine sites after a similar period of burial, have completely corroded and have reverted back to 100% mineral. This is entirely due to the presence of chlorides in seawater.

   For metals to corrode, it is essential for water and oxygen to be present. Removal of either of these will arrest the corrosion process. This helps to explain why some artefacts can be recovered after hundreds of years underground or on the seabed and still have a considerable amount of un-corroded metal left. The formation of concretions on the surface will have hindered the arrival of dissolved oxygen in the seawater and slow or even arrest the corrosion process. The heavy wrought iron guns from the *Mary Rose* were prevented from corrosion by being buried deep in the Solent silt. This prevented the ingress of oxygen to the metal surface. Artefacts buried in soils may remain uncorroded by similar effects.

Problems arise when these items are excavated and oxygen is again able to reach the metal surface. Corrosion re-commences at an alarming rate if aggressive ions like chlorides are present in the corroded layers on top of the metal surface. The aim of the conservator is to prevent this from occurring and to remove all the deleterious ions in the corrosion products. This must be carried out without destroying the original shape of the artefact and then allows the conserved artefact to go on display or in a store without suffering any further corrosive attack.

In order to fully appreciate the reasons for carrying out the conservation method selected, it is important to understand in the first instance how the metal or alloy was manufactured. From modern theories of corrosion of metals in marine environments, it is possible to predict the mode of corrosive attack that the artefact may have experienced while being buried or laying on the bottom of the ocean floor. Any adverse effect on the rate of corrosion on exposure to the atmosphere can possibly be predicted. From this knowledge, the most efficient methods of field treatments, storage conditions and conservation can be recommended.

Methods of obtaining the metal or alloy from its ores will be briefly covered here as with the possible shaping processes to produce the finished object. The corrosion behaviour of the metal while buried will also be discussed and this will be related to various methods available in the conservation of these materials.

1.1 Extraction

Metals are rarely found in their native form with gold being the main exception. They are present in the earth's crust as minerals or ores and exist in chemical combination with other elements, the most important ones being oxygen and sulfur. Hence, the most common ores are oxides and sulfides, which are mixed with gravel, limestone, sand and clay. This unwanted material is referred to as gangue. The amount of metal present in the ores varies from less than 1% to 10%. These ores are very stable and have existed for many thousands of years. It requires a lot of energy to remove the oxygen or sulfur from the metal and the best way to achieve this is to heat the ore in the presence of a reducing agent. This process is called smelting and the chief reducing agent is carbon either in the form of charcoal or coal. The active reducing agent would either be carbon or more likely in the presence of just sufficient air, carbon monoxide gas, CO.

$$2C + O_2 = 2CO \qquad (1)$$

The reducing agent will combine with the oxygen or sulfur part of the ore and free the metal. The unwanted material forms a slag, which may be in the form of some silicate or oxides of other metals found associated with the metal being extracted. The whole process can be represented by the following simple equation:

$$\text{Ore} + \text{reducing agent} + \text{heat} = \text{metal} + \text{slag} + \text{gas} \qquad (2)$$

Oxygen from the air must be prevented from mixing with the metal or else it will immediately react to produce fresh oxide. Additionally, the presence of oxygen may exhaust the reducing agent. This indicates that the design of the furnace required to heat the ore/reducing mixture, must be such that there is a reasonably large region in which no oxygen is present. This is where the reaction in Equation (2) takes place and the atmosphere is said to be reducing. The temperature at which the metal is produced varies from metal to metal and the rates of the reaction will be faster if the reactants are in the liquid phase, *i.e.* the metal and slag are molten. This will depend on the melting point of metal. Table 1 indicates the melting of some of the common metals and the temperatures required to reduce the ores to the metal.

One of the main limitations of the early metallurgists was that they could not generate particularly high temperatures in their furnaces. Metals with low-melting points such as tin, lead and copper were produced in large quantities much earlier than the higher melting point ones such as iron. In reality, the production of wrought iron was carried out virtually by solid-state reduction with

Table 1 *Typical ores from which some common metals are extracted*

Metal	Chief Ores	Chemical Formula	Reducing Temperature (°C)	Metal Melting Point (°C)
Copper	Malachite	$CuCO_3 \cdot Cu(OH)_2$	800	1083
	Chalcotite	Cu_2S		
	Chalcopyrite	$CuFeS_2$		
Lead	Galena	PbS	800	327
Tin	Cassiterite	SnO_2	600	232
Iron	Haematite	Fe_2O_3	1000	1535
	Limonite	$Fe_2O_3 \cdot 3H_2O$		
	Magnetite	Fe_3O_4		
	Siderite	$FeCO_3$		

the end products being blooms. These were solid mixtures of iron and slag from which the slag was removed by hammering. It was not until the beginning of the fourteenth century that furnace designs allowed the temperature to reach 1000°C and only in the eighteenth century when coal began to be used as fuel, could temperatures be attained to near the melting point of iron.

Lead was one of the first metals to be extracted as it melts at 327°C and is reduced by carbon just below 800°C. The main ore is galena (PbS), and may be readily smelted in a charcoal or dry wood furnace. At the top of the furnace where there is an abundant supply of oxygen (oxidizing), the sulfide is roasted to the oxide:

$$2PbS + 3O_2 = 2PbO + 2SO_2 \tag{3}$$

The conditions within the furnace change from oxidising to reducing (no oxygen) as the ore moves down the furnace. The remaining unreacted lead sulfide mixed with the oxide, acts as a strong reducing agent and converts the oxide to metallic lead as shown in Equation (4):

$$2PbO + PbS = 3Pb + SO_2 \tag{4}$$

The molten lead is run off at the bottom of the furnace into clay moulds.

The chief copper containing ores are malachite $CuCO_3 \cdot Cu(OH)_2$, chalcocite (Cu_2S) and chalcopyrite ($CuFeS_2$). The production of copper from its ores is far easier from the oxide than the sulfide-bearing ones. The malachite is mixed with charcoal and placed in a furnace with some type of bellows arrangement to increase the temperature in the furnace to 700–800°C. Under reducing conditions in the furnace, the charcoal is converted into carbon monoxide gas, which is the reducing agent at these temperatures. The ore is reduced to metal according to Equation (5):

$$CO + CuCO_3 = 2CO_2 + Cu \tag{5}$$

The copper usually remained in the furnace as a solid ingot if the temperature could not exceed the melting point of pure copper (1083°C). As furnace design improved and temperatures above this value were achievable, the metal could be tapped off as a liquid.

Sulfide-bearing ores were first roasted by placing the chalcocite or mixed copper/iron sulfide ore over burning wood in shallow cavities for up to 30 days to convert the sulfides to oxides of copper according to Equation (6):

$$Cu_2S + 2O_2 = 2CuO + SO_2 \tag{6}$$

The oxide so produced is then treated as for the oxide-bearing ores. There is very little iron found in copper obtained from chalcopyrite as the iron is more difficult to reduce than copper and ends up in the unwanted slag.

The tin-bearing ores are found in Devon and Cornwall with the chief one being cassiterite or tinstone (SnO_2). As the ore is already in the oxide form, roasting does not need to be carried out and the cassiterite is directly reduced by the carbon in the fuel (charcoal or dry wood) to produce tin. The temperature required for this reaction is approximately 600°C and the molten tin (melting point 232°C) runs out of the bottom of the furnace into clay or stone moulds. It is important that the ore and the fuel are well mixed as they are loaded into the top of the furnace in order to ensure that the reaction takes place.

The major economic ores from which iron can be extracted are haematite (Fe_2O_3), limonite ($Fe_2O_3 \cdot 3H_2O$), magnetite (Fe_3O_4) and siderite ($FeCO_3$). The extraction of the pure metal from these ores is far more difficult than it is for producing pure copper or pure tin. The reason for this is the oxygen is more strongly bonded to the oxides of iron and requires more stringent reducing conditions within the furnace to achieve this separation. The minimum temperature at which the oxide is reduced is 800°C, which is well below the melting point of iron. Another consideration is that the ore is mixed with a large amount of unwanted material, called gangue. This has to be separated from the metal by a process of slagging whereby the unwanted material is melted by combing with silica (sand) and is drained away from the iron. The melting point of this slag is approximately 1200°C.

Shaft furnaces were employed in which the ore was mixed with charcoal and the temperature was raised to approximately 1000°C by means of manually-operated bellows. This is below the melting point of iron, so reduction occurred within the solid phase. The charcoal combines with any air in the furnace to produce carbon monoxide gas as in Equation (1). The following reactions then take place between the oxides of iron and the carbon monoxide gas:

$$Fe_2O_3 + CO = 2FeO + CO_2 \tag{7}$$

$$FeO + CO = Fe + CO_2 \tag{8}$$

Reaction (7) occurs higher up the shaft while Reaction (8) nearer the hearth. The end result was a mixture of iron and slag, which was subsequently hammered while hot to produce a bloom of wrought iron. This was called the bloomery process for the production of wrought iron. The wastage of iron was very high with over 70% of available iron lost in the slag.

The introduction of mechanically-operated bellows and taller shaft furnaces allowed higher temperatures to be achieved. These were called charcoal blast furnaces, but they did require the use of an excessive amount of fuel to reach temperatures up to 1200°C. The walls of the furnace were well lagged to keep the temperatures up, while the extra height allowed the iron ore to be in contact with the carbon monoxide gas for a longer period of time. At these higher temperatures, the solid iron slowly dissolved the charcoal, which lowered the melting point of the iron. When the carbon content becomes approximately 4.5%, the metal becomes liquid at the maximum furnace temperature and forms a layer beneath the now molten slag. This molten metal has the composition of cast iron. The slag can be separated now from the cast iron by draining the top liquid layer of slag away. The slag produced by this blast furnace has much lower iron content and is a more efficient method for producing ferrous metals. The molten cast iron is run off from the furnace into blocks called pigs or into the desired shape clay moulds. The first blast furnace built of this type in England was at Sussex in 1496, and the earliest artefact of cast iron has been dated as 1509.

Wrought iron could be produced from cast iron by blowing air over the surface of molten cast iron. This caused the carbon to oxidise to form carbon dioxide gas and this was continued until all of the carbon had been removed.

The melting point of steels is only a few degrees lower than that of pure iron and it was impossible to produce these temperatures until the middle of the eighteenth century. Previous to this, it was found that wrought iron, reheated in a charcoal fire for a sufficiently long time could be made far harder than the original wrought iron. The reason was simple the carbon had diffused into the iron to form steel. Eventually, the cementation process was developed where wrought iron was packed with charcoal in stone boxes and heated to 900°C for one week. Later developments included the re-casting of cementation steel to remove the slag. This produced superior quality steel and became known as crucible cast steel and improved the fame of Sheffield watchmakers. Methods to produce a large tonnage of steel began in 1856 with the introduction of the Bessemer process and the Siemens–Martin (open hearth) in 1865. In the former process, air (oxygen) is blown through a charge of molten "pig iron" (cast iron) contained in a pear-shaped converter. The air converts the carbon to a gas as well as most of the other impurities in the cast iron. The reaction rates are so high that the temperature rises above the melting point of steel. The molten steel was subsequently cast into moulds. The process of converting cast iron to steel took only approximately 30 min, but control of the quality of the steel was difficult. The principle of the Siemens–Martin process was similar but the time taken to produce steel was 8–10 h. The quality of the steel was easier to control by this method. Both processes flourished for over 100 years until 1952 with the development of the oxygen lance process.

It is possible to produce alloys from ores. Archaeological bronze is an alloy of copper and tin and the alloys were made by mixing tin ore, (cassiterite, SnO_2) with pure copper, covering the mixture with a layer of charcoal and heating to approximately 800°C. Liquid alloy was tapped from the furnace.

Brass is an alloy of copper and zinc and the manufacture of this alloy was achieved by taking pieces of copper and mixing them with calamine ($ZnCO_3$), a zinc ore and charcoal and placing in a crucible. The crucible was heated to 950–1000°C to reduce the calamine to zinc vapour which dissolved in the pure solid copper. The temperature of the crucible was raised to melt the alloy, the temperature being dependent on the amount of zinc dissolved in the alloy. A 20% zinc alloy had a melting point of 1000°C, while a 30% alloy was slightly lower at 904°C

Silver is extracted from sulfide bearing ores of lead and copper. To be viable, the silver content should be in the order of 0.033%. During the production of "pure" lead, all the silver present in the ore, will have dissolved in the lead. The cupellation process was used to recover the silver from the lead. The silver–lead alloy was heated in hearths with bone ash to a temperature of 1000°C (melting point silver 960°C) and air blown over the surface. The lead, together with any other base metal, was oxidised while the silver remained unaffected. The oxidised lead (PbO – litharge) was skimmed off and a small button of metallic silver was left in the hearth.

1.2 Metallurgy

In the solid state, metals are crystalline, *i.e.* the atoms are arranged in a regular three-dimensional pattern with cubic structures being the most common. This accounts for the excellent mechanical properties of metals such as ductility and toughness. Ceramics and glasses have extremely complicated crystal shapes and, as a result, are very hard and brittle at room temperature. Due to their crystal structure, it is possible to form alloys of two or more metals and this can result in a considerable improvement in certain mechanical properties such as strength and hardness.

Pure metals are easy to shape by hammering, beating or bending because of their softness and ductility. Mechanical working of metals is what blacksmiths, goldsmiths and silversmiths have been doing for centuries. If this process is carried out at room temperature, it is called cold working and the hardness is increased, but the ductility is decreased by this process. There is a limit to the amount of change of dimensions that the metal can withstand by cold working, as eventually the metal will be so hard that it will crack if any further deformation is carried out. Cold working was thus a very simple way to harden a pure metal but the increase would not be sufficient to manufacture an axe head, for example. Metals could have their ductility restored by reheating them to temperatures in excess of 200°C. This is called annealing and if the shaping of metals was performed at these elevated temperatures, it was referred to as hot working. For steel on heating to red heat, it becomes soft enough for large changes of shape to be made with relatively small forces, hence the blacksmiths craft. The temperature of hot working is peculiar to each metal, with temperatures in excess of 550°C being required for copper, while the annealing temperature for lead is room temperature. This is why it is impossible to harden pure lead by cold work.

Metals and alloys that were shaped by mechanical means were called wrought alloys, with wrought iron and wrought brass being good examples of this class of materials.

For some alloys or the early impure single metals, it was virtually impossible to shape them by either cold or hot working. Provided the metal or alloy could be kept molten, it was possible to pour the molten metal into specially shaped moulds where it would eventually solidify into a similar shape to that of the mould. This is called casting and relies on the ability of a molten metal to take up the shape of vessel (mould) containing it. The moulds were made from clay or sand and were destroyed after each pouring. Casting can fairly easily produce complicated shapes and it would be very difficult if not impossible to manufacture the same item by mechanical means. This is the reason why large wrought iron split-ring guns were superseded by cast iron ones at the beginning of the sixteenth century!

If a polished piece of metal is placed in a suitable chemical (etchant), grains will be visible if viewed under an optical microscope. The size of these will vary between a few microns (μm) up to 3–4 mm in diameter. If an alloying element is added to the parent metal and can fit into the original crystal structure, the two metals are said to be soluble in one another (similar to milk in tea). Under the microscope, the grains will appear the same as the pure metal. These alloys are often referred to as single-phase alloys. The addition of alloying elements will increase the hardness without sacrificing too much loss in ductility, provided the alloying additions are able to dissolve in the crystal structure of the parent metal. Arsenic (4%) dissolved in copper (often referred to as arsenical copper alloy) had a hardness approximately double that obtained from pure copper. Pure silver was too soft for coinage so copper was added to increase the hardness and wear resistance. A typical analysis of coinage silver, which did not vary for many centuries, was 6.19% copper, 0.8% lead and 0.3% gold together with other trace impurities such as arsenic and antimony. Brass is an alloy of copper and zinc in which the zinc is completely soluble in the copper lattice up to 30 wt%. The advantages of adding zinc to copper was that the tensile strength and hardness of the resultant alloy was superior to that of pure copper. Depending upon the impurities present in the alloy, the ductility was also improved. As already stated, the term wrought brass was often applied to this type of alloy.

If the two metals that form the alloy are insoluble in one another, then they will exist as two separate phases, often in alternate layers such as observed in tin–lead alloys or cast irons, where the carbon is often found as minute tadpole like shape (flakes) adjacent to the pure iron. These types of two-phase alloys are extremely difficult if not impossible to shape by hot or cold working. Fortunately, these alloys have a melting point well below that of the parent metals and are very suitable to shape by casting into moulds. This is the reason why iron 4.5% carbon alloys were called cast irons. These alloys have two important limitations in that first, they are very brittle when subjected to impact loads, and second, their corrosion resistance is inferior to pure metals or single-phase alloys.

Metals and alloys are commonly divided into two major types, Ferrous or Non-Ferrous. The former are alloys of iron while the latter includes the remainder of the metals including copper, lead, tin, silver, gold and aluminium and alloys of these metals. Ferrous materials can be divided into three types as follows:

(1) Wrought iron
(2) Cast iron
(3) Steel

Wrought iron is essentially pure iron with particles of slag from the refining method still included in the structure. Wrought iron melts at 1535°C and is relatively soft and ductile and, therefore, is able to be shaped by such techniques as forging and hammering. As a result of these shaping processes, the slag orientates itself in the direction of working (streaks of slag) and the end result is a fibrous structure.

Cast iron is an alloy of iron and carbon with the latter ranging from 2% to 4.5%. The melting point of this alloy is between 1150°C and 1200°C, which is considerably lower than wrought iron. The carbon is found in the cast iron as graphite, which can take the shape of flakes (grey cast iron) or spheres (spheroidal or ductile cast iron). In some instances, the carbon will be in the form of iron carbide (cementite, Fe_3C) and the alloy was called white cast iron. Both forms of carbon make the cast iron a very brittle material and thus it was impossible to shape it by forging or hammering. Fortunately, due to the lower melting point of cast iron, it was possible to melt the iron and cast it into the required shape.

Steels are alloys of iron and up to 1.7% carbon, although steels are not usually found with more than 1.2% carbon. The importance of steels is that their mechanical properties are greatly influenced by their carbon content. As the carbon increases in the steel, the ductility goes down while the hardness and tensile strength go up. A further important consideration is that the hardness of steels can be dramatically increased to even higher levels by a process of heating to above ~800°C and quenching in water or some other fluid such as urine and then tempering at 200–300°C. This produces the ideal mechanical properties for cutting instruments such as knives, blades, scissors, razors, *etc.*, which require sharp cutting edges and do not blunt easily.

Some of the non-ferrous metals and alloys have already been mentioned such as copper and brass. Archaeological bronze is an alloy of copper and tin. The composition of these alloys ranges typically from 3% to 14% tin together with trace impurities such as lead and iron, depending on the chemical content of the original ores. Alloys with a tin content, up to 6%, were capable of being cast and subsequently hammered into their final shape. This is due to the tin being soluble in the copper crystal structure, which allows the alloy to be deformed at room temperature.

With higher levels of tin, the melting point of the alloy is lowered, with a corresponding increase in fluidity, thus making these alloys ideal for shaping by casting. This is the method used for producing the range of bronze cannons recovered from the *Mary Rose*. Adding zinc improved the fluidity of the alloy still further and allowed less tin to be used. This was particularly important if there was a shortage of tin or it became too expensive. These ternary alloys were the precursor to modern Admiralty Gun-Metal, which is a 88/10/2 alloy of copper/tin/zinc.

Lead was one of the first metals to be smelted and used by man. The low melting point (327°C), softness, malleability and ductility indicated the ease with which it could be cast and formed. As the annealing temperature for lead is room temperature, it is impossible to harden pure lead by cold work. Alloying is the only way to increase the hardness of lead. Lead was used for weights (density 11.35 g cm^{-3}), sheeting, piping, net sinkers, etc.

The major alloy of tin recovered from archaeological sites is pewter. This can be divided into those containing lead and lead-free alloys. The former could have a lead content ranging from 67% (equivalent to plumbers solder) down to 15%. The French in Elizabethan times kept the lead of their wine goblets to below 18% as above this, the wine would become tainted! As the lead and tin are insoluble in one another, they are classed as a two-phase alloy and articles could only be manufactured by casting. The lead-free pewter was invariably an alloy of tin with a small amount of copper (0.5–7% for pewter recovered from the *Mary Rose*). The copper dissolved in the tin crystal structure resulted in a single-phase structure, which was considerably harder than pure tin. Hence this class of pewter could be subjected to a limited amount of mechanical working to achieve the final shape.

2   CORROSION

The aqueous corrosion of metals is due to an electrochemical cell being formed between two different metals in electrical contact (galvanic corrosion) or two separate metallic phases present on a single metal surface. Consider a strip of copper (copper electrode) joined to a strip of iron (iron electrode) by an insulated wire and immersed in sea-water (approximately 3 wt% sodium chloride). The iron will start to dissolve (corrode as Fe^{2+}) and release electrons according to Equation (9)

$$Fe = Fe^{++} + 2e \qquad (9)$$

In corrosion science, the iron electrode is referred to as the anode and this is always where corrosion takes place.

The electrons travel through the wire to the copper electrode surface where they react with dissolved oxygen gas in the seawater to form hydroxyl ions according to Equation (10)

$$O_2 + 4e + 2H_2O = 4OH^- \qquad (10)$$

The copper electrode is called the cathode and no corrosion occurs at this electrode. The electrical circuit is completed by the seawater electrolyte, which carries a charge by the movement of ions through the solution. This would be via the movement of chloride and sodium ions present in the seawater.

Single metals and alloys such as wrought iron, bronze and lead will corrode even though they are not joined to a different metal or alloy. In reality, anodes and cathodes are set up on the surface of the metal. This will be due to the different phases present in the alloy, *e.g.* iron (ferrite) will be the anode and graphite the cathode in a cast iron, impurities such as sulfides present in wrought iron and variation in the copper/zinc ratio in adjacent grains in a brass alloy. The reactions taking place on the anode and cathode will be the same except the anode may be a different metal if the object was not iron. With a copper artefact, for example, the anode reaction will be:

$$Cu = Cu^{++} + 2e \qquad (11)$$

where Cu^{++} denotes that copper is now present in the solution. The cathode reaction will be the same as Equation (10) for all metals and alloys in natural environments such as seawater, freshwater or soils. Cathodes and anodes will also form on metal surfaces if moisture films condense on a metal surface on a damp day or if the humidity inside a museum rises above a critical value.

From the above, the following five factors are required for corrosion to occur:

1. Anode – where metal dissolution occurs, *i.e.* corrosion.
2. Cathode – where the electrons are consumed by an electrode reaction. The cathode is protected.

3. Electrolyte – to complete the electrical circuit.
4. Electrical contact between anode and cathode – to allow the transfer of electrons from the anode to the cathode.
5. Cathode reactant – to use the electrons formed at the anode. Dissolved oxygen is the most common one in archaeological situations.

If any of these components are absent, the metal will not corrode. For example, sachets of silica gel are put into display cabinets to prevent corrosion by stopping an electrolyte from condensing on the metal surfaces. Museums, nowadays, use more refined methods of humidity control for cabinets.

This also explains the remarkable state of preservation of numerous artefacts that have remained undisturbed for many hundreds of years prior to excavation. A good example of this is the wrought iron strip ring guns recovered from the *Mary Rose*. These had lain deeply buried in silt on the floor of the Solent for 440 years and were excavated in very good condition. The reason for this is that the silt prevented oxygen from reaching the metal surface: no oxygen, no cathode reactant and no corrosion. In other instances, the artefact has been covered with thick layers of concretion, which have prevented, once more, oxygen reaching the metal surface. On removal of the concretion, corrosion will re-commence as oxygen can now reach the metal surface and act as cathode reactant (Equation (10)). This highlights the importance of treating the artefact immediately it has been excavated. Even if the concretion is not removed on excavation, oxygen can reach the cathode sites as this layer is very brittle and readily cracks to allow air (oxygen) to the metal surface.

The metal ions formed at the anode (Equation (9)) will react with the hydroxyl ion formed at the cathode (Equation (10)), or any other ions present in the electrolyte to form metal compounds. The ferrous ion formed in Equation (9) will form a compound $FeO \cdot OH$, which is red rust. While in the presence of low levels of oxygen in the electrolyte, black magnetite (Fe_3O_4) will be the main corrosion product. On excavating ferrous artefacts, black corrosion products are often observed but these change to red rust on exposure to the air after several hours. Seawater contains 3 wt% sodium chloride and ferrous chloride ($FeCl_2$) may form as well as ferrous carbonate ($FeCO_3$) due to the hardness of salts found in this natural environment. Over 35 different iron compounds have been identified on ferrous artefacts recovered from soil, fresh- and sea-water sites. Copper and its alloys produce aesthetically-pleasing corrosion products ranging from red, purple and black through to green and blue. The famous green patina on copper is a basic copper carbonate either malachite ($Cu_2(OH)_2CO_3$) or azurite ($Cu_3(OH)_2(CO_3)_2$).

Once the corrosion products have formed on the metal surface, the subsequent rate of corrosion will depend on whether these compounds can block the arrival of oxygen to the cathodic sites or the dissolution of metal from the anode sites. Lead artefacts form relatively insoluble compounds as corrosion products in marine environments such as $PbSO_4$, $PbCl_2$ and $PbCO_3$. These form adherent films on the metal surface that insulate the metal from the electrolyte and prevent further attack. This results in lead artefacts being recovered from marine sites in very good condition even after burial for over 400 years, *e.g.* lead shot and weights from the *Mary Rose*.

Tin and pewter artefacts with low lead contents will form a non-protective SnO_2 layer on the surface in well-aerated seawater environments. Hence, artefacts recovered from the surface of the seabed, will be almost completely mineralised. The shape of the artefact may still be maintained by the tin oxide, but it will be very fragile and must be handled with great care. In anaerobic conditions, however, a very protective sulfide film will form, which virtually inhibits further corrosion. Artefacts recovered from these sites are usually in excellent condition and require minimum conservation.

Silver items recovered from marine sites are often completely mineralised due to the non-protective nature of the corrosion products formed in both aerobic and anaerobic sites. The corrosion product is either silver chloride (AgCl) or silver sulfide (Ag_2S). All silver artefacts recovered from the *Mary Rose* were found to be in very poor condition.

As mentioned above for pewter objects, the corrosion products may maintain the shape of the object even though no metal is left. One of the best examples of this is observed in cast irons. The iron phase (ferrite) corrodes to form the same corrosion products as already stated above. The graphite flakes in the cast iron are inert and trap the corrosion products and the shape is maintained. This is called graphitisation of cast iron and almost all cast iron artefacts recovered from archaeological sites have graphitised layers on their surface.

Chloride-containing compounds are, in general, less protective than the corresponding oxides, hydroxides, carbonates, etc. The chloride ions are said to be very deleterious as they will rapidly corrode most of the metals with the possible exception of gold when the artefact is freshly exposed to the atmosphere. There are several reasons for this excessive rate of corrosion. The first is that chlorides will readily dissolve into any moisture films that condense on the artefact during display or in storage and increase the conductivity of the water. Ionic conduction is easier through the electrolyte and results in a higher flow of current in the corrosion cell. The second reason is that chloride-containing compounds in the corrosion products tend to be more soluble than oxides or hydroxides, which are the predominant compounds formed in the absence of chlorides. Liquid corrosion products will not hinder the arrival of oxygen to the metal surface and the cathode reaction is less impeded. The chloride-containing compounds are also quite deliquescent and readily attract water. To maintain completely dry surfaces on the metal artefact, it is necessary to keep the humidity as low as 16% RH for some chloride-containing compounds. This is difficult to achieve in most circumstances in museums or storage rooms. Finally, corrosion reactions involving chlorides can produce acids such as hydrochloric acid. These reactions may take place in crevices or pits on the metal surface and the pH can drop from 8.1 (seawater) down to 2.5 at the base of a pit. Metals will invariably corrode faster in acids than in neutral electrolytes, and hence corrosion will be accelerated at the base of these pits. From the above, it is apparent that if the artefact is to survive long after it has been excavated, the removal of the deleterious chloride is essential. This is one of the main tasks in the conservation of metals as once these have been removed, the metal is said to be stabilised.

Dissolution of the chlorides from the corrosion products is an essential part of the conservation process. It is essential that the artefact is immersed in an electrolyte that will not corrode the metal any further, while this dissolution is taking place. Corrosion scientists have developed redox potential – pH diagrams from thermodynamics in order to predict the most stable form of the metal. These diagrams are divided into three zones. Where metal ions are the most stable phase, this is classed as a zone of corrosion. If the metal itself is the most stable species, this is said to be the zone of immunity. The third zone is where solid metal compounds such as oxides, hydroxides, etc, are the most stable and may form a protective layer over the metal surface. This zone is termed passivity and the metal will not corrode as long as this film forms a protective barrier. The thickness of this passive layer may only be approximately 10 nm thick but as long as it covers the entire metal surface, it will prevent further corrosion.

A typical E-pH diagram for iron is illustrated with a diagram in Figure 1. The zones where Fe^{++} and Fe^{+++} are stable are the zones of corrosion, while the zone of immunity is where the metal, Fe, is the stable phase. The zones where the oxides Fe_2O_3 and Fe_3O_4 are stable are termed passivity and the iron will not corrode. By measuring the pH and the steady potential of the metal against a reference electrode, it is possible to determine the zone where the metal is situated in any given environment. Iron in seawater (pH ~8) and freshwater (pH ~7) will give a point on the E-pH diagram where Fe^{++} is most stable *i.e.* iron will corrode. As one is only too well aware, steels and cast irons readily corrode in these environments. If the pH was raised to above 9 by the addition of alkali such as sodium hydroxide to freshwater, the point on the diagram would be in the zone of passivity where Fe_2O_3 is the most stable phase. The iron will remain passive and not corrode as long as the pH is maintained at this value. Using alkalis such as sodium hydroxide or sodium carbonate in solutions, above pH 9, is one of the most accepted methods for the removal of chloride ions from ferrous artifacts without corroding any remaining metal. Inspection of the relevant E-pH diagram for the metal/water system will indicate the range of pH which may be used for the soaking of chlorides from the artefact. A pH of 9 or above would be a disaster for aluminum as it would corrode

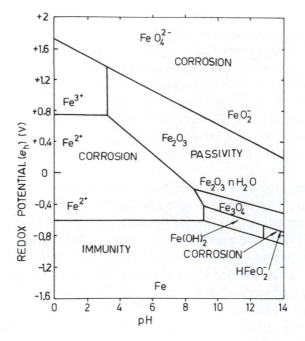

Figure 1 *Pourbaix diagram for iron; in an environment with a dissolved iron concentration of 10^{-6} g·atoms L^{-1} at a temperature of 25°C*
(Modified by Wranglen, 1985.)

rapidly in these strong alkali solutions. Aluminum artefacts, such as old engines, would have to be treated in a solution with a pH in the range 4–8. Care must be exercised in the use of E-pH diagrams as there are some limitations to the use of these in predicting the corrosion behaviour of metals. As they are based on thermodynamics, kinetic considerations are neglected. Prior to selecting a pH range from the E-pH diagram for the metal that one wishes to conserve, it is prudent to test metal coupons in the solution to ensure that a passive film will form and that there is no corrosion of the uncorroded metal or its corrosion products. If this was the case, there will be no artefact left in the solution after a few weeks of soaking!

3   FIELD TREATMENT AND TRANSPORTATION

Metal artefacts immersed in seawater for long periods of time such as over hundreds of years, may become completely corroded and 100% mineralised. In certain instances this is the case but in others, the metals or alloys have only exhibited minimal amounts of corrosion. Even ferrous artefacts recovered from the same site may experience vastly different corrosion rates.

Figure 2 *Concretion and mud on a split ring gun recovered from the Mary Rose*

As stated in the corrosion section, metals immersed in seawater have all the requirements for a corrosion cell to be set up on the surface. Why is there this enormous difference in corrosion rate? The reason is due to the exclusion or reduction in dissolved oxygen reaching the metal surface. The oxygen supply could be restricted due to the artefact being deeply buried in the sea-bed. The large wrought iron split ring guns from the *Mary Rose* are a good example of this and this is well illustrated in Figure 2, *e.g.* where the bore of the gun recovered from the Solent is full of mud from the sea-bed. Another reason for dissolved oxygen not being able to diffuse to the metal surface is that the artefact is covered with a thick layer of concretion (calcium carbonate). These concretions can be from 0.5 to 15 cm thick. These arise from hardness salts in the seawater changing to solid carbonate during the initial corrosion reactions on the metal surface and/or the settling of barnacles, mussels, *etc.*, on the metal surface. These die to form a thick carbonate scale. With some metals and alloys, the corrosion products themselves may limit the arrival of dissolved oxygen to the metal surface. The lead artefacts recovered from the *Mary Rose* are a good example of this where, even after immersion for hundreds of years in

Figure 3 *Artefacts recovered from the seabed*

the Solent, they were in very good condition and were protected by a thin 1–3 mm of corrosion products.

Once the artefacts are brought up from the sea-bed as illustrated in Figure 3, oxygen can reach the metal surface and corrosion can recommence due to damage or removal of the protective layers. Artefacts require some form of immediate treatment, therefore, to prevent deterioration while awaiting permanent conservation. In addition, slow atmospheric drying produces a concentrated salt solution in the pores or cracks beneath concretions, *etc.*, which is conducive to rapid corrosion. The role of the chloride ion in the corrosion process has already been discussed and as the concentration of chloride in the remaining electrolyte increases so does the rate of corrosion. Cast iron is very much prone to this form of attack. A final problem is that the corrosion products will slowly dry out and lose their water of crystallisation. This involves a reduction in volume of the corrosion products, which may lead to cracking and exfoliation of the corroded surfaces. The shape of the artefact may be lost or seriously damaged and dissolved oxygen can now easily reach the metal surface with a dramatic increase in corrosion rate.

The first priority is to keep the artefact wet immediately it is brought to the surface. If the object is very small and shows little evidence of corrosion, it may be placed in an airtight container along with a desiccant such as silica gel. In the absence of nearby conservation laboratories, totally immerse the artefact in seawater or preferably freshwater. Corrosion of any remaining metal will occur but none of the problems mentioned above will commence. Very large artefacts such as the cannons from the *Mary Rose*, could be covered with wet sawdust, damp cloths or plastic wrapping in an attempt to maintain water levels around the metal surface.

If there is a conservation laboratory nearby, immerse the artefact straight away in an electrolyte, which passivates any exposed metal. The pH can be ascertained from the relevant E-pH diagram for the metal concerned and examples for ferrous artefacts would be 0.5 M sodium hydroxide, 0.2 M sodium sesquicarbonate or 0.5 M sodium carbonate.

For transportation of the artefact it is essential to maintain treatment initiated in the field. If the items cannot be moved in their storage tanks, they must be packed as mentioned for the large artefacts above, *i.e.* use sawdust, *etc.* If by the smell of the corrosion products (presence of hydrogen sulfide gas), microbial corrosion of the artefact is suspected, it would be advisable to add a biocide in order to minimise this type of attack until it can be treated in the laboratory.

4 CONSERVATION

The main aims for the conservation of metals are as follows:

1. Arrest the corrosion process;
2. Remove chloride ions from the corrosion products;
3. Leave the shape of the artefact unaltered; and
4. The metallurgical structure of the artefact remains unaltered.

The removal of the chloride ion is the most essential, but the conservator must consider the other three aims when considering what method to use. For example, there are several proprietary solutions on the market or ones that can be made up in a laboratory, which will dissolve all the corrosion products but not the underlying metal. This method is not suitable if the artefact has thick layers of corrosion products on the surface as the shape will be lost. Even worse, if the artefact was just composed of solid-corrosion products (completely mineralised), there would be nothing left after immersion in these solutions! This method is only suitable for those artefacts recovered with thin layers of corrosion products on their surface.

A relatively simple method is to dissolve out the chloride ions by immersion in a suitable solvent. Water has been used with the water being changed every month until no further chlorides are detected. This can take up to 5 years for marine artefacts with high levels of chloride buried within deep rust layers. Moreover, the metal will continue to corrode, while the artefact is immersed in the water for this length of time. By altering the pH of the solution it may be possible to dissolve out the chlorides without corroding the metal. This is achieved by forming a thin, passive film approximately $10 \, \text{nm} \, (\sim 10^{-9} \, \text{m})$ thick on the exposed metal. For example, wrought iron will form a passive film in a solution above pH 9. Hence sodium hydroxide, sodium carbonate or ammonium hydroxide would be suitable for this class of artefacts. The disadvantage of these solutions is that they still take a long time to remove all the chlorides and one is not absolutely certain that all the chlorides have been removed even after a period of 5 years.

For artefacts recovered from marine sites, they are often covered in concretions. These are hard layers of calcareous deposit derived from decaying shells of aquatic animals (*e.g.* barnacles, mussels, *etc.*) or hardness salts present in seawater. The latter is present as soluble bicarbonate ions (HCO_3^-) in sea or fresh water. At cathodic sites on the metal surface, there is a rise in the local pH due to the production of OH^- during the reduction of dissolved oxygen gas (see Equation (10) in corrosion section). This results in the precipitation of solid calcium carbonate ($CaCO_3$) scale on the cathodic sites according to the following reaction:

$$Ca^{++} + HCO_3^- + OH^- \rightarrow CaCO_3 + H_2O \qquad (11)$$

The concretions arising from the two sources mentioned above, may be up to 15 cm thick and must be removed prior to any conservation method being carried out. This has to be done carefully as it is very easy to cause severe damage to the artefact particularly if it is in a very fragile state.

There are several techniques employed by conservators, but the most popular ones are those using mechanical methods. It is advisable to carry this out when the artefacts are moderately dry just prior to carrying out the conservation process. The common methods in this class are by use of chisels, dentist drills or shot blasting using bauxite, glass beads, magnesium carbonate or just air.

Chemical methods have been utilised where the concretion is dissolved in a suitable solvent. Solutions such as 1–2 M nitric acid, 2.5 M phosphoric acid and 2–3 M hydrochloric acid have been used, but there is a severe danger of the metal dissolving if the artefact is not removed at the exact moment when all the concretions is being dissolved. Inhibitors could be added to the acids but one is never certain of the exact composition of the metal beneath the concretion. Each metal or alloy will have its own inhibitor so it is very easy to add the wrong inhibitor.

Finally, thermal methods such as placing the artefact in a hydrogen furnace at 400°C has been successfully employed on ferrous artefacts recovered from the *Mary Rose*. Any remaining concretion on the surface was present as a fine powder after conservation and was just gently brushed off the surface. An alternative thermal method involved placing the artefact in an inert atmosphere at 1066°C followed by quenching. The procedure is repeated until all the concretion has spalled from the metal.

Figure 4 *Concretion recovered from seabed with no visible indication of metal object within*

Sometimes the artefact may be brought to the surface from the sea-bed and the shape of the artefact may be readily apparent even though it is covered in a thick layer of concretion. Even though the shape is obvious, the artefact may be completely mineralised or only a small amount of uncorroded metal left. In the worse case scenario, there may be a void beneath the concretion layers. Alternatively, the shape of the artefact may be impossible to ascertain as the concretion is just a solid lump (Figure 4). In all these cases it is advisable to use electromagnetic radiation to examine what is beneath the concretion. X-rays and γ-rays have been used for this purpose. These have a shorter wavelength and higher energy than does visible light and are thus able to pass through material which is totally opaque to visible light. The extent to which X-rays and γ-rays can pass through a material depends upon the thickness of the arte-fact and the atomic weights of the elements in the artefact. The light elements such as C, N, O and H (found in the concretion) are virtually transparent, while the commonly found metals such as Cu, Fe, Sn are medium absorbers but Pb is a strong absorber. The artefact is irradiated from one side with penetrating electromagnetic radiation and a photographic film is placed on the opposite surface. A shadowgraph of the artefact is obtained. Scales, corrosion prod-ucts, cracks, pores, *etc.*, absorb little radiation and show up as darker regions on the film. Solid metal, if present, shows up as very light areas. It is possible to use a TV screen instead of the photographic film.

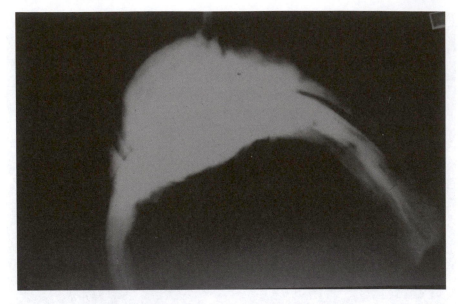

Figure 5 *Radiograph of concretion showing shape of the object which was later identified as a rudder pintle*

X-rays are not very good for thick objects and the equipment is not portable. Radioisotopes such as Co^{60}, Ir^{192} and Cs^{137} in the form of a wire and sealed in an aluminium capsule are used as a source of γ-rays. The whole unit is stored in a lead or tungsten alloy, which is opaque to γ-rays. Radioisotopes are portable and it is possible to distinguish between two metals beneath a concretion. The subsequent radiograph of the undistinguished lump in Figure 4 (which turned out to be a rudder pintle), is shown in Figure 5. Once the shape of the underlying metal is revealed, the conservator is able to delicately remove the concretion without destroying the shape of the object as shown in Figure 6.

If the extent of corrosion of the artefact while buried is uncertain, a radiograph may be taken to ascertain the amount of metal left uncorroded. This was often helpful in deciding the most suitable method for subsequent conservation.

4.1 Acid Pickling

In the metal finishing, the use of acids for the removal of corrosion products prior to coating is widely practiced. The acids used depend on the metal or alloy being treated as it is essential that once the corrosion products have been removed, there is no attack on the exposed metal surface. For use in conservation the aim is to selectively dissolve the rust and leave the parent metal unaffected. Thus 5 M HCl plus hexamine as corrosion inhibitor has been used for

Figure 6 *Careful removal of concretion reveals the iron rudder pintle*

ferrous artefacts, as has 2.5 M H_3PO_4, which leaves a film of ferric phosphate on exposed iron. A 1–2% HCl solution has been used on lead artefacts. This dissolves concretions and corrosion products but does not attack the underlying metal.

The main drawback of this method is that it removes surface layers and thus alters the shape of the artefact. Indeed, if the artefact is completely mineralised, the complete artefact will dissolve and there will be nothing left! This contravenes one of the important aims of conservation in that the artefact's shape must be unaltered. Another serious shortcoming is that traces of acid may be trapped in crevices and cracks in the artefact, which will continue to rapidly corrode the metal after conservation during storage or on display. Overall, apart from lead-based artefacts, this method is not recommended in the majority of cases.

As an alternative to the use of acids, sequestering agents have been employed to dissolve the corrosion products without attacking the parent metal. The most effective formulations are based on the derivative of ethylene diamine–tetra acetic acid (EDTA). Lead artefacts from the *Mary Rose* were cleaned in a 10% solution of this compound. The use of EDTA is not recommended for cast iron as the graphite flakes embedded within the corrosion products are also dissolved. As with the use of acids, the shape of the artefact is altered if the corrosion layers are very thick and it is also difficult to wash out all the solutions from cracks, crevices and pores in the artefact after cleaning.

4.2   Aqueous Washing

The principle of this method of conservation is to immerse the artefact in a tank containing a suitable solution. The chloride ion dissolves from the rust film into the solution that is changed, initially, every week and subsequently every month. The chloride content of the solution is analyzed at the end of each changeover. The process is continued until there is no more chloride detected. At this point, the artefact is deemed to be conserved. This can take up to 5 years for marine artefacts with high levels of chlorides buried within deep rust layers. Even after this length of time, one is not absolutely certain that all the deleterious ions have been removed from the rust/metal interface.

A relatively simple method is to dissolve out the chloride by immersion in distilled or de-mineralised water. Most of the common metals and alloys will continue to corrode in this solution although at a slower rate due to the lower conductivity of these solutions compared to seawater. This can be serious if the artefact is to be immersed for periods up to 5 years, but has been employed if the immersion times are relatively short. A small copper alloy sheave recovered from the *Mary Rose* was soaked in running demineralised water for 27 days and the chloride levels dropped from 106 to 10 ppm over this period.

By using boiled water, the dissolved oxygen is expelled and hence, there should be no corrosion as the cathode reactant has been eliminated from the electrolyte. Unless the boiled water is kept in sealed containers, air (oxygen) will slowly dissolve into the water and corrosion of the metal or alloy will re-commence. As an alternative, using hot demineralised or distilled water will reduce the concentration of dissolved oxygen and hence corrosion, but this must be counter-balanced by the rise in reaction rates with temperature. In open conservation tanks, a temperature of 70°C is required to notice a significant reduction in rates of corrosion of metals. Small copper alloy artefacts from the *Mary Rose* were treated in this way using water at 80°C for 30 days. At the end of this period, the chloride levels in the water dropped to below 1 ppm.

Some conservators have alternated between tanks of boiled and cold water for their artefacts. They claimed that the expansion and contraction of the artefact, will assist in the removal of the deeply-buried chlorides from the rust layers. One must be careful that this does not cause the rust to spall off the underlying metal due to the difference in expansion coefficients between the two classes of materials.

By altering the pH of the solution, it may be possible to dissolve out the chlorides without corroding the metal. This is achieved by the formation of a thin, passive film, approximately 10 nm thick on the exposed uncorroded metal. The pH required to passivate any given metal or alloy can be determined by inspection of the relevant E-pH diagram. For wrought iron, a passive film will form above a pH value of 9.5. This would be a disaster for aluminium artefacts as these will corrode the metal at an alarming rate. A pH in the range 4–8 is required to form a passive film on aluminium.

Sodium hydroxide is a very common solution used for ferrous artefacts. A concentration of 0.5 M will give a pH of approximately 10.5 that is well in the passivity region for this class of materials. The problem is that if any ferrous or ferric chloride type compounds are present within the corrosion products, these may react with the hydroxide to produce solid sodium chloride within the pores in the rust film (Equation (12)).

$$FeCl_3 + 3NaOH = Fe(OH)_3 + 3NaCl \qquad (12)$$

The iron oxide precipitates within the rust film and traps the sodium chloride in the corrosion products. This is a serious problem when using a combination of aqueous and thermal methods of conservation as will be discussed later.

Sodium carbonate is another widely used solution for the conservation of ferrous artefacts. This maintains the pH in the zone of passivity (pH > 9.5) and is also said to act as an anodic inhibitor. A concentration of 3 g dm^{-3} (0.2 M) was used in the conservation of *Holland 1* in Gosport Submarine Museum. It is advisable to use the sesquicarbonate form as it is far easier to dissolve than the simple carbonate. The use of this solution is said to avoid formation of solid sodium chloride in the rust.

The treatment of small ferrous artefacts in a 0.05 M lithium hydroxide dissolved in methanol or ethanol has its advocates, particularly in France. The chlorides present in the rust layers react with lithium hydroxide to form lithium chloride that dissolves in the alcohol phase. Any of the hydroxide left on the metal surface combines with any carbon dioxide to form a solution with pH above 9.5, which maintains any exposed metal in the passive region. Hence, this solution is claimed to cause no corrosion of the underlying metal. The real disadvantage of this solution is that any lithium chloride left on the surface of the artefact is very hygroscopic. Water will form on the surface at a relative humidity above 15% RH and corrosion of the metal will take place. Humidity levels below 15% RH are very difficult to maintain in display cabinets or in storage and is one of the main reasons why this solution has not been more widely employed.

Alkaline sulfite reduction was developed in an attempt to reduce the time to fully conserve the artefact. The pH of the solution is maintained in the passive region by the use of sodium hydroxide (0.5 M) while at the same time, the sodium sulfite (0.5 M) in the mixture will slowly reduce the red rust (FeO · OH) to magnetite (Fe_3O_4).

$$3FeO \cdot OH + H^+ + e^- = Fe_3O_4 + 2H_2O \qquad (13)$$

Figure 7 *Holland 1 at the submarine museum Gosport, England prior to conservation*

This reduction involves a 30% decrease in volume, which makes the rust layers more porous and allows the solution to reach the deeply-buried chlorides. The conservation must be carried out in sealed containers as ingress of oxygen into the solution would convert the sulfite to sulfate, and hence no reduction would take place. Some conservators have heated the solutions up to 60°C to speed up the reduction rates, and hence the removal of chloride ions. The two major problems with this particular formulation are that the chloride analysis is rather difficult and is not very efficient for artefacts covered in thick layers of corrosion products.

The conservation of the first submarine of the British Royal Navy, *Holland 1*, was carried out by aqueous washing method (Figure 7). The submarine was launched in October 1901. She had a crew of nine and weighed 105 tons with a length of 19 m and a maximum diameter of 3.7 m at her widest point. A four-cylinder petrol engine producing 160 Hp and weighing 4 tons, gave the submarine a top surface speed of 7.4 knots. Underwater propulsion was provided by a 74 Hp electric motor powered by 60 lead acid battery cells each weighing 108.5 kg. This gave a maximum underwater speed of 7 knots. There were two ballast tanks into which water was admitted to increase the vessel's weight and allow her to dive to a maximum of 30 m. However, the normal operating depth was 15 m, which took approximately 40 min to reach from the surface. *Holland 1* saw 12 years service before she was superseded by more advanced and powerful craft and she was sold for scrap in 1913. While being towed to the breaker's yard, she sank off the Plymouth coast and remained on the seabed until her discovery by a Royal Navy minesweeper in April 1981.

Holland 1 was raised in September 1981 and taken to dry dock in Devonport where she was cleaned with high-pressure water to remove all the concretion and marine growth. To facilitate transportation to the Royal Naval Submarine Museum at Gosport, Hampshire, the submarine was cut into three sections. On arrival at Gosport, the submarine was welded back together and the exterior of the hull painted with a commercial anti-corrosion paint, which was claimed to give long-term protection. The interior of the submarine was painted with a decorative white paint and the engine and its ancillary parts with a decorative green finish. After 11 years of exposure to the marine environment of the Museum, the exterior paint had virtually disappeared and the hull was badly corroding. The interior was also showing signs of severe corrosion particularly at the seams produced by the overlapping of the steel plates of the hull and around the fasteners such as nuts and bolts. Parts of the engine had completely mineralised and had been severely damaged by the constant stream of visitors being shown around the interior. The exterior of the hull was once more pressure cleaned to remove the unsightly corrosion products.

The original specification for the thickness of steel plates used in the manufacture of the hull was 11 mm. Core samples were taken by trepanning and the thickness of the metal remaining was measured. The results ranged from 9.0 down to 2.5 mm, with the bow section having experienced the most severe corrosion. In certain areas, the steel had completely corroded with gaping holes clearly visible. This illustrates the wide variation in corrosion rates experienced by the submarine as it lay for some 70 years on the seabed.

Metallurgical examination of the core samples indicated that the hull was manufactured from a mild steel with a carbon content in the range 0.1–0.15% with a hardness value of 152 VPN. The carbon content of the steel rivets was a little higher at 0.18% a hardness value of 239 VPN. The 160 Hp engine has been manufactured from grey cast iron. Corrosion products removed from the interior of the hull and from internal crevices such as the overlapping of steel plates, gave a chloride value of 2.21%. This is of a similar magnitude to that found by other conservators on ferrous artefacts recovered from marine sites. With this level of chloride in the rust, it is not surprising that corrosion of the steel plate had continued underneath the paint film.

Besides the ferrous materials, several large components of the submarine were manufactured from non-ferrous materials. These include the coning tower, torpedo tube and propeller. These were all manufactured from the same material, a two phase, α/β brass (60% copper–40% zinc alloy) with lead additions to improve the machinability properties of the material.

As the submarine was too fragile to move, the only practical conservation method was to wash the vessel *in situ*. A tank was constructed from glass fibre – reinforced polyester resin laminate. The structure was prefabricated in sections away from the museum and assembled on site with no disturbance to

Figure 8 *A fibre glass panel being positioned to form a cell around the submarine*

the submarine. Figure 8 shows one of the sections being lifted by crane around the submarine.

The lower edges of the tank were bonded to a load-bearing concrete dish, which had been previously laid beneath the submarine and also incorporated the steel cradle, which had supported the hull while on display at the museum for the previous 13 years. In order to strengthen the tank, steel ribs were inserted within the laminate at regular intervals along its length. Cross braces of 15 mm steel cables were passed through the hull of the submarine to give further reinforcement when the tank was full of the electrolyte. The total volume of liquid contained within the tank was $820,000\,dm^3$ and this posed a problem in the choice of electrolyte. Figure 9 shows the completed tank for the aqueous washing of the submarine. Permission had to be sought to dispose of this quantity of solution into public sewers. In addition, in the event of a leak developing in the tank, the solution could flow into Portsmouth Harbour as the submarine was only approximately 50 m from the harbour wall. Sodium hydroxide was deemed to be unsafe by the local authority and sodium carbonate was the solution chosen to meet both conservation and environmental requirements. To ensure that the concentration of carbonate and chloride remain uniform throughout the tank, a re-circulating pumping system was installed. The electrolyte in the tanks returned to the pump via a stainless steel mixing pump. The emptying of the tank to remove the dissolved chloride ions took place via a valve located in the floor of the vessel, which exits directly into the public drainage system.

Figure 9 *The complete cell formed around Holland 1 for aqueous washing of the chloride ions from the submarine*

Prior to the assembly of the glass fibre-reinforced polyester resin laminate tank, as many of the non-ferrous fittings as feasibly possible were removed. The stripping of the paint coatings on the interior of the submarine and on the cast iron engine did pose problems, particularly on the latter, as it was severely graphitised and brittle. Conventional grit blasting would have damaged the original shape of the engine to say nothing about the dust problem within the confined space of the interior of the submarine. This was solved by using the *Spongejet* system, which has mild abrasives encased in sponge particles. These composite particles are directed at the surface by high-pressure hoses. The sponge absorbs the paint plus any contaminants on the surface thus minimising the amount of dust generated in the atmosphere within the submarine. The very gentle nature of this abrasive cleaning process causes very little damage to the underlying metal.

One of the main considerations in the conservation of the submarine was the galvanic corrosion between the body of the submarine (ferrous), the coning tower, torpedo tube and propeller (brass) and the copper windings of the electric motor. In seawater, the copper and brass would be protected while the steel body would corrode. In 5% sodium carbonate solution, there is a polarity reversal with the steel being protected and the copper-based materials corroding. If traces of copper ions were to escape into Portsmouth Harbour through a leak in the tank for example, severe biological problems would ensue as copper ions are very toxic to marine life. Research showed that by dropping the concentration of sodium carbonate to 3%, this reversal could be prevented. Reducing the sodium carbonate to even lower levels decreased the pH to below the range where the steel remains passive. Reference electrodes were positioned throughout the submarine to monitor the potentials of both the steel and brass in order to ensure that the steel remained just in the passive region and the brass did not corrode with release of toxic copper ions.

The tank was filled up with the required amount of sodium carbonate solution and regular chloride analysis was carried out, first at weekly intervals and subsequently at approximately monthly intervals. Solution changes were not carried out as frequently as desired as the cost of chemicals to fill a 820,000 dm³ was rather excessive. After 3 years, no more chlorides were detected in the electrolyte. The next stage of conservation was the construction of a viewing gallery where the relative humidity could be maintained below 35% at all times even with large numbers of visitors passing through. This ensured that no moisture films formed on the metal surfaces of the submarine. The submarine was coated with an industrial wax. This was relatively hard wearing, but did allow simple visual detection of any subsequent sign of corrosion of the ferrous and copper based components of the submarine.

4.3   Electrolytic Conservation

In this method, the artefact is immersed in a tank of electrolyte in which the metal will not corrode, *i.e.* it remains passive. An electrolytic cell is formed with the artefact being the cathode with an inert anode. A small dc current is passed between the two electrodes and the corrosion products on the surface undergo a reduction process to a different compound. In the case of iron, the following takes place on the cathode surface.

$$3FeOOH + e = Fe_3O_4 + OH^- + H_2O \qquad (14)$$

For one mole of Fe, the volume of FeOOH (red rust) is 21.3 cm⁻³ and the volume of Fe_3O_4 (magnetite) is 14.9 cm⁻³. This involves a 30% reduction in volume, which allows the electrolyte to penetrate more easily through the corrosion products and reach the deeply-buried chloride ions. In addition, the negatively-charged chloride ion migrates due to the influence of the electric field, from the cathode (artefact) toward the anode. This speeds up the rate of chloride ion removal and decreases the time for conservation.

The anodes that have been used include stainless steels, mild steel, lead and platinised titanium, while typical electrolytes for ferrous materials have been 0.5 M sodium hydroxide, 0.2 M sodium carbonate, 0.5 M sodium sesquicarbonate and tap water. For bronze cannons recovered from the *Mary Rose*, both sodium hydroxide and sodium carbonate electrolytes were employed while pewter artefacts (plates) from the same ship were treated in similar electrolytes or in a 0.5% solution of EDTA as a sodium salt in alkaline solution.

The important precaution to be considered in this method is the evolution of hydrogen gas on the surface of the artefact, which is the cathode in the cell. For this reason the current densities must be kept to low values. Typical current densities that have been used range from 300 to 2000 mA m^{-2} depending on the metal/alloy being protected. Careful monitoring of the potential of the cathode by use of reference electrodes must be employed in order to maintain the potential of the artefact below the hydrogen evolution potential. This can be obtained from the relevant E-pH diagram. If hydrogen gas is evolved, it can cause layers of corrosion products to spall off the surface and significantly alter the shape of the artefact. For situations involving artefacts with high-chloride levels in the rust layers, there is the strong possibility that chlorine gas may be liberated at the anode. This can be a problem in enclosed spaces, particularly if this gas reaches high levels of concentration. Despite these reservations, this is quite an effective process for the conservation of metals and the conservation time can vary from 2 to 33 months depending upon the original chloride content buried within the corrosion products.

The only surviving First World War ship belonging to the Royal Navy was brought into Portsmouth harbour in 1990 and then into dry dock beside *HMS Victory* in 1997 for the purpose of conservation. At the beginning of the First World War, the Admiralty embarked on a very extensive ship-building programme. These included vessels called Monitors, which were to provide shore bombardment. In order to be able to sail close to the shore, the 6000 tons Monitors had shallow drafts of 1.8–3 m (6–10 ft) and two revolving 15 in gun turrets. For use in very shallow waters and river estuaries, smaller vessels of only 600 tons with a draft of only 1.2 m (4 ft) and two 6 in revolving guns placed fore and aft were ordered in March 1915. One of these was M.33 which was built in 2 months and launched in May 1915. She had a top speed of 9.6 knots but had to be towed for most of her first journey to the Aegean and the Dardenelles as her engines were only designed for 3 days continuous use. M.33 remained in active service for the remainder of the war and then served in campaigns in Russia during 1919. Later in 1925, she was converted to a minesweeper and re-named *Minerva* and finally converted to a staff office and workshop during the Second World War.

HMS Minerva was finally "paid off" in 1984 and eventually brought by Hampshire County Council and berthed in No 1 Basin in Portsmouth Naval Base in 1990 and transferred into dry dock in 1997. (Figure 10) An ultrasonic survey was conducted over strategic parts of the ship in order to determine the extent of corrosion of the steel plate. The original thickness of the metal was 12.5 mm and the results showed that the average metal remaining over the ship was 8 mm. Near the stern of the ship around the steering gear and propellers, only 3 mm of the original metal plate was left. The reason for the high-metal

Figure 10 *M.33 in dry dock in Portsmouth dockyard*

loss was due to the turbulence in the surrounding seawater while the ship was underway.

Metallurgical analysis revealed that the steel plates were normalized plain carbon steel with a carbon content of 0.06–0.08% and a hardness of 172 VPN. Chloride analysis of the rust films gave an average value of 0.33% which may appear to be relatively low but could be explained by the continual washing of the metal surface by rain water which assisted in the prevention of build up of high chloride concentrations in the rust films. Samples taken from the interior of the ship gave considerably higher values for chloride content in the corrosion products. The chloride levels were 8–10 times those found on the exterior, with the magazine compartment giving values of 3.93%. This highlights the benefits gained by the washing of metal surfaces by rainwater. The magazine compartment contained a primitive refrigeration system, which worked by means of a brine tank. Leakage of brine from this could have contributed to the high chloride levels discovered in the rust layers recovered from this area.

Once in dry dock, the outside of the ship was pressure-washed with mains water to remove marine growths attached to the steel plates as well as the loosely adherent corrosion products. This was repeated several times to assist in the removal of chloride ions from the rust layers. Approximately 11 tons of debris were removed from the external structure of the ship by this process. Several sections of the steel plates were found to have very thin areas less than 1 mm

Figure 11 *Stainless steel anodes used in the electrolytic conservation of M33*

thick, which did not show up on the ultrasonic survey. New plates were inserted if the area of damage was excessive or a patch weld applied if the area of damage was not excessive. These restoration areas were documented for future reference.

The interior of the ship was conserved by the electrolytic method as this had the highest levels of chloride in the rust layers. Sections of the hull were segregated from each other and ensured that they were all watertight. A section was flooded with water and the pH adjusted to 9.5 by adding sodium carbonate. This maintained the steel in the passive region and improved the conductivity of the electrolyte. The latter reduces the power requirements of the dc supply used to electrolyse the cell. The anodes were manufactured from 316 stainless steel (austenitic stainless) mesh (Figure 11). By using the mesh, gases such as oxygen and chlorine, which may be produced on these surfaces, were able to escape and the higher grade of corrosion-resistant stainless steel considerably reduced the pitting of the anodes. The chloride content of the electrolyte was monitored daily at first and then weekly. If the chloride content became too high, the electrolyte was changed in order to reduce the likelihood of pitting corrosion. The current density used throughout the conservation programme was approximately $350\,\text{mA m}^{-2}$. Once no more chloride was observed to be coming out of the rust layers, the electrolysis was terminated. Each section was left with the sodium carbonate solution in it to keep the steel in the passive region. Once all the sections had been conserved, they were drained, pressure-washed, dried and painted with a temporary coating until the future of the ship has been determined. The coating chosen was a rust converter *Ferrozinc* (HMG Paints) that converts any remaining red rust into a protective layer, which does not allow the corrosion reactions to take place on its surface.

4.4 Hydrogen Reduction

Several thermal methods have been used in an effort to conserve metal arte-facts. For those that have been completely mineralised, heating in a furnace at 800°C for 1–12 h was said to drive off all the volatile chlorides within the corrosion products. The deeply-buried chlorides would still have been trapped within the corrosion products and several forms of chlorides would not have sublimed under these conditions. A modification of this technique involved soaking the artefacts in ammonium carbamate/ammonium hydroxide solution prior to heating in a furnace under vacuum. This soaking was said to convert all the chlorides into ammonium chloride, which is volatile at elevated temperatures. This would be true if all the chlorides were originally in the form of sodium chloride but this is not the case for most metals and alloys. Another variation on this theme was to pack wood charcoal around the artefact prior to placing it in a furnace at 800°C. The wood creates a slightly reducing atmosphere, which assists in the removal of the chloride ion. As air was not excluded from the furnace, it is unlikely to have reduced the oxides of iron to any great extent.

The hydrogen reduction conservation process was first employed in Sweden in 1964 for ferrous artefacts recovered from the Swedish warship, the *Vasa*. The method was further developed at Portsmouth to treat the large number of finds recovered from the Solent and land-based archaeological sites within the Wessex region. The principle of the process is to heat the artefact in an atmosphere of hydrogen in order to sublime off the volatile chlorides and at the same time reduce the oxides, hydroxides, chlorides and eventually to the metallic state. The volume change associated with the reduction of the iron compounds is sufficiently high to enable the release of deeply-buried chlorides particularly at the metal/corrosion product interface.

The furnace at Portsmouth was installed in the Conservation Department of the City Museum in 1975. The design of the furnace was influenced by the considerable number of large cannons recovered from the Solent and in particular, from the *Mary Rose*. It was a vertical retort furnace with a cylindrical bell type retort 2.5 m high by 0.7 m in diameter made of *Nimonic*, a heat resistant alloy. Once loaded with the artefacts, the retort was wheeled into an electrically-heated furnace with three banks of independently-controlled electric heating elements in order to produce an even temperature distribution within the retort. The reducing atmosphere was obtained by passing ammonia over a nickel catalyst at 850°C to produce an atmosphere of 75% hydrogen and 25% nitrogen. This mixture of gases was then fed into the retort at a slightly positive pressure and out at the bottom of the retort and exhaust for analysis and extraction of harmful constituents.

Initially, ferrous artefacts were stored in 2% sodium hydroxide to prevent any further corrosion prior to treatment in the hydrogen furnace. It was subsequently found out that the sodium hydroxide converted some of the chlorides in the rust layers to sodium chloride. This was only very slowly removed in the retort as it does not sublime at the operating temperatures or get reduced by the hydrogen to release the deleterious chloride ion. Ammonia-based electrolytes were subsequently used for pre-storage prior to the artefact undergoing hydrogen reduction. The chlorides are now converted to ammonium chloride, which sublimes off at temperatures in excess of 520°C and undergoes reduction in the hydrogen atmosphere present within the retort.

After removal from the soaking tanks, the concretions were removed from the artefacts. For large items such as the guns, this was mostly carried out by mechanical methods, using chisels or scalpels. Light concretions on small artefacts were not touched as they were either removed during conservation or were easily brushed off after removal from the furnace. The final pre-treatment were to make quite sure that the barrel of the guns was clean so that the atmosphere within the furnace could freely reach all parts.

Once cleaned, the artefact, if it was a gun–barrel, was secured onto a stainless steel base (Figure 12) and the *Nimonic* alloy retort clamped down over it so as to ensure an airtight seal.

Small artefacts were placed onto stainless steel mesh trays and also fixed into the retort so as to give maximum circulation of the furnace atmosphere during the conservation process. The retort was wheeled into the electrically-heated furnace purged and with nitrogen until the last traces of oxygen were removed. The cracked ammonia replaced the nitrogen as the atmosphere within the retort and the temperature of the furnace was allowed to rise slowly to the operating temperature over a 24 h period. For cast iron and most artefacts recovered from marine sites, this would be 850°C. This was to ensure that the artefact was not subjected to any significant thermal shock. The temperature would remain at this value until the chlorides in the exhaust gases coming out of the furnace were below 10 ppm as detected by Draeger tubes. This could be anything from 100 to 200 h depending upon the original levels of chloride within the rust layers. The cracked ammonia was replaced by nitrogen in the furnace and cooled down over a 48 h period. The artefacts were removed from the retort and given a very light brush to remove any powdery deposits left on the surface. The

Figure 12 *Loading a gun onto the retort of the hydrogen furnace*

original corrosion products were now converted mostly into a mixture of magnetite and metallic iron. The reactions that have taken place may be as follows:

$$2FeOOH \rightarrow Fe_2O_3 + H_2O \qquad (15)$$

Figure 13 *Similar caronades from HMS Pomone before and after conservation in the hydrogen furnace*

(red rust)

$$Fe_2O_3 \rightarrow Fe_3O_4 \tag{16}$$

(magnetite)

$$Fe_2O_3 + 3H_2 \rightarrow 2Fe + 3H_2O \tag{17}$$

Chloride-based iron compounds either sublime at the operating temperatures or reduced in the presence of hydrogen to form hydrogen chloride. A typical reaction would be

$$FeCl_2 + H_2 \rightarrow Fe + 2HCl \tag{18}$$

This magnetite/iron mixture is very friable and must be consolidated immediately if the shape of the artefact is to be maintained. The large guns recovered from the Solent were placed in tanks of low-viscosity epoxy or polyester resin where impregnation occurred. Figure 13 shows two similar guns from HMS Pomone, one before, and the other after, treatment in the hydrogen furnace.

Small items from the furnace were impregnated with the same resins in a vacuum chamber.

The major criticism of this method of conservation is that the metallurgical structure of the artefact has been altered by heating the artefact to elevated temperatures. For this reason, relatively few artefacts with derived microstructures such as knives, axes, *etc.*, were treated in the furnace at Portsmouth. If hydrogen treatment was the method selected for these, the operating temperature was reduced to 350°C and the process time extended in order to minimise the alteration to the microstructure. Nevertheless, this method is one of the best and quickest methods for the removal of the chloride ions. Guns from the *Mary Rose* were fully conserved within 2–3 weeks and even after exposure to the atmospheric conditions for nearly 30 years, they still show no sign of breakdown.

5   STABILISATION AND CONSOLIDATION

Almost all washing and electrolytic conservation treatments leave the artefact in a wet condition. This moisture has to be removed if further corrosion is to be prevented. For small artefacts, drying in an oven is very common, while for large ones such as cannons, hot air blasts or infrared lamps are employed.

In order to minimise the surface corrosion after conservation, dewatering fluids can be used. These consist of compounds which displace water from the surface, *e.g.* butanol, plus a corrosion inhibitor. With copper and its alloys, problems arise if any chlorides are still present in the corrosion products after conservation and the humidity in storage or on display rises above approximately 40% RH. Serious deterioration of the artefact can take place by the hydrolysis of the chloride compound, cuprous chloride. These hydrated compounds occupy a far larger volume than the original ones and erupt through the corrosion layers often as light green spots. This is given the name bronze disease and can lead to complete destruction of the artefact if left unchecked. In order to counteract this problem, the majority of copper-based artefacts from the *Mary Rose* were soaked after washing in a 1–3% alcoholic solution of benzotriazole (BTA) for up to 2 weeks. BTA is a well-known corrosion inhibitor for copper and its alloys. This functions by forming a linear polymer of copper and BTA on the metal surface. It is not perfectly clear, however, why this inhibitor stops hydrolysis of the cuprous chloride in the corrosion products. It is possible that the BTA reacts with the chloride compound to form a protective polymer film. Alternatively, a complex could be formed between the cuprous chloride and BTA, which renders the chloride unaffected by humid conditions.

Many artefacts are in a fragile and delicate state following conservation. The highly friable and porous surface layers are impregnated with a suitable polymer to achieve an acceptable mechanical strength. Thermosetting resins have been successfully used on the wrought iron guns from the *Mary Rose*. These polymers have a high degree of cross-linking that imparts good mechanical strength to these resins. Typical examples in this group include epoxies and polyester resins.

Thermoplastics in the form of a lacquer have poorer mechanical strength than the thermosetting resins but are more easily removed, should this prove necessary. Polyvinyl acetate and polyurethane are good examples of this class of consolidants, particularly on wrought iron artefacts. Many of the copper and copper-based alloys, such as bronzes and brasses recovered from the *Mary Rose*, were consolidated with a solution of acrylic resin dissolved in toluene (*Incralac*).

Waxes are probably the most extensively employed consolidants because of their versatility. The artefact could be immersed in molten wax, which solidifies within the pores of the corrosion layers. Naturally-occurring beeswax is a common example, while microcrystalline waxes are now formulated to meet the requirements of the conservator such as hardness, colour, melting point, *etc.* The surface of the wax can now be further modified by the application of graphite powder to give an appearance similar to that of a wrought iron finish.

6 CONCLUSIONS

In treatment of metal artefacts, the conservation scientist must have knowledge of metallurgy, corrosion and chemistry. Marine artefacts recovered are invariably covered in concretions but a radiograph helps to ascertain the shape of the metal underneath prior to removal of this layer. Identification of the metal or alloy follows, together with mode of manufacture and any thermal treatments used to improve mechanical properties. This information, together with knowledge of the burial environment, will allow the conservation scientist to predict the nature of the corrosion products enveloping the remaining metal and indicate the most suitable method of conservation. Finally, once the object is conserved, post-conservation treatments must be carried out or metal corrosion will soon recommence.

REFERENCES AND FURTHER READING

C. Pearson, *Conservation of Marine Archaeological Objects*, Butterworths, London, 1987.

D.A. Scott, *Metallography of Ancient Objects*, Summer School Press, Institute of Archaeology, London, 1987.

D.A. Scott, J. Podany and B.B. Considine, *Ancient and Historic Metals: Conservation and Scientific Research*, Getty Conservation Institute, Singapore, 1994.

D. Watkinson, *First Aid for Finds*, UKIC Archaeology Section, London, 1987.

G. Wranglen, *An Introduction to Corrosion and Protection of Metals*, Chapman & Hall, London, 1985.

CHAPTER 7

Glass and Ceramics

HANNELORE RÖMICH

Fraunhofer-Institut für Silicatforschung (ISC), Bronnbach 28,
97877 Wertheim-Bronnbach, Germany

1 INTRODUCTION

Glass and ceramics are two materials that appear rather different at first sight: glass is appreciated through its transparency, whereas ceramics are opaque materials, more related to earth than to light (see Figures 1 and 2). Even if we look at their structure and degradation mechanism, there are fundamental differences: glass is a dense material reacting only on the surface, with water as the principal aggressive agent, whereas ceramic objects are often endangered by salt precipitation within their porous structure.

The links between the two are to be seen first in their chemical composition, both being related to silicates, and second in their production process, emphasising their importance for art and archaeology. Ceramics, the older

Figure 1 *A stained glass window from Cologne Cathedral, 16th century*
(Picture provided by Dombauhuette Koeln, Germany.)

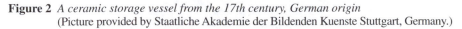

Figure 2 *A ceramic storage vessel from the 17th century, German origin*
(Picture provided by Staatliche Akademie der Bildenden Kuenste Stuttgart, Germany.)

invention, and glass, which followed later, are important indicators of the technological development of mankind. Improvements of their basic but sophisticated manufacturing techniques have defined milestones in the history of technology. Moreover, both materials reflect the creative potential and the artistic expression of craftsmen throughout our history. Apart from objects of daily use, decorative art and architectural elements have also been produced during the centuries and testify to the needs and achievements of a specific culture. Thus glass and ceramics have common links in chemistry and in history and they are often considered together, not only in the field of conservation but also in science and technology.

2   DETERIORATION AND CONSERVATION OF GLASS

2.1   History of Glassmaking

Before dealing with man-made glasses, their natural analogues should be mentioned. Glasses occur in nature as a result of the rapid cooling of silicate melts (magmas) as they come into contact with water, ice or the atmosphere. Silicate melts may originate from magmatic activity or from meteoritic impact and exhibit a broad range of chemical composition (from 30 to 80 mass% SiO_2, up to 20 mass% MgO and CaO). Obsidian, as one of the most popular representatives of natural glasses, has been used by man as an extraordinarily effective sharp-edged tool from the earliest times of our archaeological record.

The cradle of glassmaking is difficult to define, both concerning the date and the place of birth of this new invention. Northern Mesopotamia at a time prior to 2500 BC is an estimation shared by many experts. At first, decorative objects and glass beads had been hand-formed or cast using simple tools and finished by abrading. Later on, glass was moulded or pressed to form vessels.

Around 50 BC glassblowing was invented, which is considered a technological milestone in the production of glass.

Throughout the Roman Empire, new centres for the manufacturing of vessel glass were established. Glass was no longer considered as an item of luxury, being used to produce containers for storage of goods. After the decline of the Roman Empire, glassmaking experienced a slump, both in quality and quantity. Natron, as a source for soda, an important raw material for the production of glass previously supplied through Roman trade routes, became difficult to obtain. By the end of the first millennium, potash, derived from the ashes of burnt trees, replaced soda as fluxing agent in northern Europe. This change in raw materials defines an important shift in chemical composition, from soda-lime silicate glasses to potash-lime silicate glasses. The difference in durability of the two types of glasses and the consequences of conservation will be discussed later.

The use of glass for windows probably has its origin in Roman times. From the 9th century onwards, however, it gained rapidly in importance and reached its perfection in the Gothic period. Pieces of coloured glass, often stained, etched, engraved or decorated with black paint (trace lines and half tones) were incorporated in a framework of lead clamps to facilitate artistic expression. Stained glass windows are among the most precious art objects of medieval times. Many examples are still to be found in their original setting, in ecclesiastical cathedrals, for example, in Cologne, Reims, Chartres, York or Canterbury. Another possibility of using glass as an architectural element is to embed small square glass pieces in cement or mortar to create pictures on the wall or on the floor. Famous examples of these mosaics from the Roman and Byzantine period are found in Rome, Ravenna, Venice or Constantinople.

This short overview cannot consider other important applications of glass in the creation of art objects, such as mirrors, reverse paintings on glass, enamels (glass fused on copper, silver or gold) and jewellery.

2.2   Chemical Composition, Structure, and Physical Properties

The structure of glass is a consequence of its chemical composition and thus the raw materials used for its production. This chapter is limited to the discussion of glass types relevant to conservation, neglecting specific developments in modern applications.

Crystalline silica (SiO_2, quartz) consists of a strong network of silicon and oxygen. Each silicon atom is surrounded regularly by four oxygen atoms. Two oxygen atoms are shared by each of the adjacent tetrahedral units. In contrast to the regular lattice of pure silica, glass has no regular structure with a long-range order.

Apart from SiO_2, other inorganic oxides, such as phosphorus oxide, can act as network formers. Alkaline or alkaline earth oxides have a different effect,

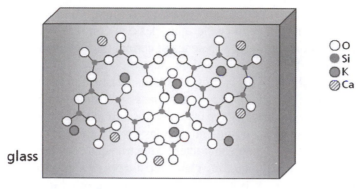

Figure 3 *Scheme of the structure of a potash-lime silicate glass. (Note: the fourth bond of silicon is out of the plane and not represented in the two-dimensional scheme)* (After Scholze, 1988.)

when reacting with silica to form a glass: the silicon–oxygen network is broken up (Figure 3). The ionic bonds between these network modifier cations and non-bridging oxygen anions from the network are less strong than the links within the network. Therefore, cations, such as potassium and calcium ions, are able to migrate within the structure, leading to a lower chemical durability. Furthermore, there is a third group of oxides (*e.g.* of lead, aluminium, magnesium) that can be incorporated into the glass structure in both ways, as network modifier or as network former.

Sand (SiO_2) was the main component for the production of ancient glass. Lime ($CaCO_3$) or magnesium carbonate ($MgCO_3$) and soda (Na_2CO_3) or potash (K_2CO_3) were added to reduce the melting temperature and to facilitate the production process. For soda- or potash-lime silicate glasses, these elements represent more than 90 wt% within the complex composition. Nevertheless, most historic glasses contain up to 30 different components that are present only in minor quantities or as trace elements. Some of them have not been added on purpose, but are present as impurities contained in the main components. Others have been chosen as colouring agents. It should be stressed that the colour of a glass is determined not only by the amount of a specific transition metal oxide but also by its oxidation state, which can be regulated during the production process by controlling the atmosphere in the furnace. Non-transparent but opaque glasses can be obtained by the presence of opacifiers, which are mainly tin, lead or antimony compounds.

In spite of some general rules for dating glasses by their chemical composition, there are many uncertainties due to the fact that the glass production of each period exhibited a broad variety in composition, although some recipes remained the same over the centuries. Nevertheless, archaeometric studies can be based, for example, on the ratio of selected major compounds (*e.g.* high lime/low alkali for 16th century German glass, in contrast to Roman glasses with a low-lime content), on the presence of certain trace elements (*e.g.* Ni, Zn) or on the isotope distribution in lead glasses.

A number of physical properties are characteristic for each type of glass composition: the viscosity of the molten glass determines the working range during production, the thermal expansion connected to the transition point (T_g) marks the transition of the liquid and the glassy state, the optical properties in relation to the refractive index characterise the transparency for light. Besides density, hardness and brittleness are characteristic physical properties. The most important chemical property for glass is its chemical durability, which is the key to degradation and described in the next section.

2.3   Degradation Mechanisms: Basic Reactions in Water

The chemical degradation of glass is initiated by water attack on the surface. The principal reactions can be explained by investigating the alteration of freshly-prepared glass in aqueous solutions. For modern glass compositions, a large number of studies have been carried out to explore the chemical durability of container glass or nuclear waste encapsulations. Their major conclusions are also relevant for historic glass. The term corrosion, originally referring to the oxidation of metals, is frequently used as a synonym for the degradation of glass.

For glass in contact with water, the degradation mechanism is clearly dominated by the pH value of the liquid. In acidic media, water and hydronium ions (H_3O^+) migrate into the glass to replace positively-charged ions of alkaline or alkaline earth elements, which are leached out of the glass. This leads to the formation of a hydrated layer, also called "gel layer" or "depleted layer", with a composition significantly different from the bulk glass (high content of Si, low content of residual Ca, Na, K) (see Figure 4).

At high pH values (in general above pH 9), another type of mechanism controls the degradation of glass: hydroxide ions attack the bridging Si—O—Si bonds, which leads to dissolution of the glass network.

Both reaction mechanisms, the ion exchange and the network dissolution, compete with each other in natural conditions. In the laboratory, those conditions can be manipulated to clearly favour one mechanism over the other.

- Glass damaged by leaching develops a depleted layer, which may not be visible under the microscope if it is limited to a few nanometers in thickness. If the leaching increases and the thickness of the hydrated layer reaches several micrometers, a characteristic pattern of interconnected cracks is developed (craquellée).

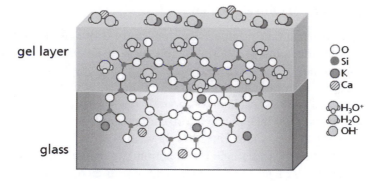

Figure 4 *Scheme of a potash-lime silicate glass after acidic attack on the surface: a chemically altered hydrated layer has formed, covered by crystalline corrosion products (Note: the fourth bond of silicon is out of the plane and not represented in the two-dimensional scheme)*

- Glass dissolution leads to a dull appearance of the glass with small dots or short cracks that are detectable only under the microscope. The thickness of the glass plate or the grains reduces with time, up to complete dissolution in extreme cases.

The reactions can be quantified even before visible damage occurs by analysing the components dissolved in the solution (preferably by ICP–AES) or by characterising the altered glass surface (with surface-sensitive techniques such as X-ray photoelectron spectroscopy (XPS), and Infrared-spectroscopy (IR)).

The rate of degradation heavily depends on the glass composition. As a general rule, the higher the percentage of silica, the more stable is the glass. Since sodium ions have stronger bonds within the network than potassium ions, the durability of sodium silicate glasses is significantly higher in comparison to potassium-rich glasses. However, sodium and potassium interact with each other in mixed-alkali glasses, which further complicates the prediction if both are present. In principle, the durability of glasses can be estimated with several theoretical approaches. These models provide good predictions for modern glasses. Historic glasses, however, have a rather complex composition. Furthermore, apart from the chemical composition, other factors such as the surface roughness, the thermal history and the production process, as well as the presence of inhomogeneities have an influence on the chemical durability. Once again, for historic glasses, these parameters complicate the prediction of their degradation with time. Consequently, model glasses with medieval compositions or with an even higher content of potassium oxide have been created to facilitate systematic research on the degradation of historic glasses.

2.4 Degradation of Objects Indoors

In a solution, the reaction products are constantly diffusing away from the glass, in contrast to natural weathering in the museum environment, where the alkaline compounds accumulate on the surface. Water droplets appear due to the hygroscopic nature of the degradation products, which is described as "weeping glasses". If the leached compounds react with air pollutants, several secondary reactions take place to form, for example, carbonates that may precipitate as crystals on the surface. In showcases without ventilation, specific pollutants such as acetic acid may accumulate leading to the formation of acetates, which are again hygroscopic and appear as droplets. Apart from the formation of crystals or droplets, the fracturing of the surface is a symptom of decay, which is visible even without microscope and often described as "crizzling". Thicker layers tend to flake away, leading to a loss of the original substance.

All examples mentioned above demonstrate that even glasses that have never or rarely been in contact with liquid water can suffer from deterioration. Above a certain limit in relative humidity, the glass surface attracts water molecules that form a film on the surface and are ready to initiate ion-exchange reactions.

Apart from chemical degradation, there are examples of glasses in museums with signs of damage connected to their former use. Many of those examples are damaged by scratches that provide prominent spots for further corrosion.

When looking at a specific museum collection, a large number of objects are in good condition, with a variable number of "sick" glasses among them. According to a survey of the National Museum of Scotland, for example, around 400 out of 2000 objects of the collection showed signs of deterioration. These glasses require special treatments and are therefore the focus of conservation research.

2.5 Weathering of Stained Glass Windows

Many stained glass windows are still set in their original surrounding and are thus exposed to outdoor weathering (if not recently fitted with protective glazing), which leads to a variety of degradation phenomena. The impact of the environment on medieval glasses leads first of all to the formation of a hydrated layer, exhibiting a crack pattern if a certain thickness has been reached. In addition, the formation of crystalline encrustations is detected frequently: leached K and Ca ions react with air pollutants to form carbonates, nitrates and mostly sulfates. Soluble compounds are washed away by rainwater. Thus, the crust analysed nowadays on medieval glasses consists mainly of gypsum $(CaSO_4 \cdot 2H_2O)$, which is the less soluble weathering product. Figure 5 shows

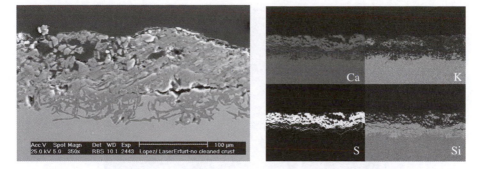

Figure 5 *Medieval stained glass window from Erfurt (Germany) with a dense crust of weathering products, a cross section, SEM picture (left) and element mapping (EDX) (right); (Note: elements are specified in the picture (Ca, K, S, Si))*

the cross section of a fragment from Erfurt Cathedral (Germany): in the SEM picture it can be seen that the degradation layer has a total thickness up to 200 μm and that this layer is not homogeneous, with many micro-cracks penetrating deep into the glass. The leached layer and the corrosion crust can be distinguished only with an additional element mapping (with energy dispersive X-ray analysis (EDX)). The layer adjacent to the bulk glass has a high Si-content and is thus defined as the "gel layer" (or "depleted layer"). The crust on top is rich in Ca and S, therefore, identified as gypsum (or weathering crust). In the sample above, both layers are mixed within a broad transition area.

Other types of glasses tend to develop different symptoms of decay: the formation of pits is related to droplets of leached compounds accumulated on the surface or to inhomogeneities in the glass. Other case studies have connected pitting with the attack of microorganisms. Local pits can be quite large (up to 3 mm in diameter). Smaller pits can accumulate and cover the whole surface of a fragment, leading to the loss of the original surface. Another remarkable phenomenon of degradation is limited to manganese-containing glasses: due to oxidation processes underneath the surface, glasses that were almost transparent turn brown.

Potassium-rich medieval glasses are very sensitive to corrosive attack. Only a few examples of glass segments with an even and almost unaffected surface are documented. Heavily corroded glasses have lost up to one-third of their original thickness. In contrast to the potassium-rich medieval windows, the soda lime silicate glasses of the 19th century are more stable and show only minor signs of degradation. Microorganisms may have settled on the surface and may have produced a bio-layer that reduces the transparency of the glass and has to be removed. Furthermore, for 19th century windows, the degradation of paint is considered to be the most serious damage phenomenon that poses serious conservation problems (Figure 6).

Figure 6 *Glass panel with damaged paint, the paint lines are partly lost (19th century, Netherlands)*

Figure 7 *Fragment of an archaeological vessel glass damaged by "iridescence": a network of micro-cracks is covering the entire surface, which has already partly flaked off*

2.6 Degradation of Archaeological Glasses

In general, Roman glass (sodium rich) is quite stable. The only sign of alteration after hundreds of years in the soil is often the formation of a thin altered surface layer. For less durable glasses this layer can reach several hundreds of micrometers, exhibiting "iridescence" (rainbow-like colouration), if several thin layers of altered glass are superimposed (see Figure 7). Surfaces of archaeological glasses may appear dull and pitted, with brown spots or dark stains. Enamel-like surface layers render the glass into a completely opaque material (Figure 8). Although the surface appears smooth, with no crystal deposits, the degradation layer is thick (leaving only a minor core of bulk glass uncorroded)

Figure 8 *Archaeological glass, medieval, with enamel-like surface layer, an overview*

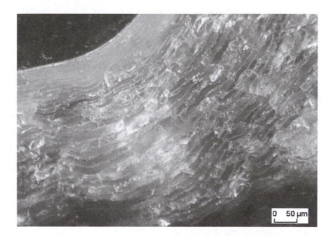

Figure 9 *Archaeological glass, medieval, with enamel-like surface layer; optical microscopy of the cross-section, showing parallel laminated layers*

and exhibits a laminated structure, due to precipitation processes. The fluctuations in chemical composition within this surface layer can be visualised in a cross-section (Figure 9). The degradation mechanism of buried glasses is complicated by the fact that transition metal ions from the ground may migrate into the altered glass.

It is difficult to estimate whether the archaeological record stored in our museums is representative of what was produced and used originally. The finds have gone through various selection processes conducted by archaeologists, restorers and curators. Highly sensitive glasses, especially in waterlogged environments, may have degraded completely and have never been identified as glass during the excavation.

2.7 Conservation Materials: Coatings, Consolidants, Adhesives

The most effective preservation method for corrosion-sensitive glasses should aim at the direct protection of the surface by sealing it with a coating against the impact of the environment. Organic natural and synthetic polymers (such as waxes, epoxy resins, polyurethanes, acrylates) have been used for this purpose in the past. However, the coatings have failed in many cases due to their poor adhesive properties for weathered-glass surfaces and their insufficient aging properties (see Figure 10). Although an improvement was proposed based on the application of silane coupling agents, modern conservation practice favours preventive conservation to avoid the long-term risk related to the application of protective coatings.

Whereas protective coatings are meant to seal the surface against further weathering, consolidants are applied to strengthen porous structures of endangered paint layers on glass (see Figure 6) or to deal with fragile archaeological glass surfaces (Figure 7). In these cases, a consolidation might be necessary to preserve the original design and thus the authenticity of the object. A risk assessment is required for each individual object to justify a treatment with a polymer, since the general requirement for the reversibility of any conservation treatment is not met in the case of fragile, porous surfaces. At present, several cellulose derivatives, epoxy resins, silanes, silicones, heteropolysiloxanes and vinyl polymers are under investigation. Paraloid B72 is the most commonly used polymer for paint consolidation, being an acrylic resin, applied in various organic solvents (such as toluene) in a broad range of concentrations (3 up to 30 mass%), depending on the purpose of the treatment. Consolidation under reduced pressure might be necessary to improve the penetration of the

0   1 mm

Figure 10 *Archaeological glass with coating: after several years of exposure indoors, the polymer lost its adhesion to the glass support*

polymer into the pores and cracks. The application of the polymer in solution with a brush or a pipette is common, followed by the removal of excess material with a solvent.

Glass, as a fragile material, often needs repair by edge bonding of broken pieces together. The best visual appearance for transparent, colourless glasses is achieved when the refractive indices of glass and adhesive match, in order to avoid reflections from the breaks. For practical reasons, epoxy resins with varying viscosity and curing time, commercially available under trade names such as Epotec, Ablebond, Araldite, Fynebond or HYXTAL, are frequently used in glass conservation. For sensitive glasses, where epoxy resins tend to provide a strong joint, which is difficult to remove after aging, weaker bonds created by cellulose nitrates or acrylates are preferred.

The adhesion of polymers to glass is best if the surface is clean. This becomes a problem when setting up the conservation strategy: consolidation before cleaning has to be favoured, for example, if the distinction between dirt and paint is difficult or for gilded surfaces. Cleaning before the application of a polymer is preferred, *e.g.* when broken pieces need strong joints.

Polymers in glass conservation may also be applied for filling gaps and modelling missing areas in broken objects.

2.8 Restoration and Cleaning

The restoration and conservation strategy for any kind of glass object has to be based on preliminary investigations, carried out by visual inspection and under the optical microscope (in transmitting and reflecting light). Heavily degraded objects or pieces with surface decoration (paint layers or engravings) will need more careful treatments than stable, unaltered sheet glass. All initial examinations should be part of the overall recording and documentation procedure.

The surface structure, and especially its roughness, is a decisive parameter for glass weathering. Therefore, the selection of an appropriate cleaning process has to take into account not only its effectiveness, but also the potential damage for the object.

In accordance with the general conservation principle of minimum intervention, the main objective is to conserve the glass, and not to recover transparency, through removal of corrosion products and deposits. Only in exceptional circumstances, therefore, may weathering layers be removed to increase the transparency of the glass or to support its interpretation. In any case, damage to the hydrated layer must be avoided: this layer is considered to be the "skin" of the glass, which protects it from further attack.

A variety of "dry" methods are at the disposal of the conservator to remove dirt and other solid deposits from the surface: soft bristle brushes may be sufficient for the removal of porous encrustations, whereas more severely

Figure 11 *Scratches on a stained glass from Cologne Cathedral, resulting from a previous cleaning process*

adhering crusts may require a treatment with glass-fibre brushes, scalpels or air-abrasive techniques, bearing a higher risk of damage for sensitive glasses. Scratches on the glass surface are prominent areas enhancing degradation, which can be detected by crystal growth, even occurring years after the damage was initiated (Figure 11).

"Wet" cleaning is preferred for several cleaning problems: sound glass can be washed with water and, even for more sensitive glasses, water is a commonly used solvent, providing it is used with care, *e.g.* when removing hygroscopic droplets from "weeping" glasses. Water may also be necessary to desalinate finds from marine sites. Complexing agents like ethylenediamine tetra-acetic acid (EDTA) or its sodium salts, as well as special gels containing ammonium carbonate, are highly effective for cleaning a gypsum crust from stained glass windows. However, the risk of further leaching of calcium ions from the glass limits the applicability of these chemicals. Ultrasonic bath treatment is extremely dangerous, when paint layers are present underneath the dirt. Organic solvents, used to remove organic polymers originating from previous conservation treatments, are rated harmless for glass. As an exception, acetone has to be singled out, as it might have the potential to dry out hydrated layers on the surface and thus induce flaking. Most recently, lasers have been investigated for the cleaning of historic glass, but no general recommendation for this technique can be given at this stage of research.

2.9   Conditions for Storage and Display

Water is the most influential parameter directing the environmental attack on glass. Therefore, the optimisation of relative humidity during storage and display is the most important concern for preventive conservation of sensitive glass objects.

For freshly-prepared glasses or glasses with a rather stable composition, as well as objects not affected by previous degradation, the rate of degradation below 35% relative humidity is very low (in the absence of acid pollutants). Therefore, these objects are protected best in a dry atmosphere. In contrast, glasses with a hydrated layer on the surface, built up during previous exposure to atmospheric weathering or to humid burial environments, require storage at higher values of relative humidity. Recommendations in the literature vary between 35–40% RH (for Egyptian faience), 42% RH (for excavated glasses), 37–42% RH (glass collection of the Veste Coburg in Germany) and 50% RH (for crizzled glasses at the Corning Museum, USA). Dry conditions provoke dehydration of these surfaces, followed by the formation of micro-cracks and peeling effects. In any case, rapid changes of relative humidity have to be strictly avoided to allow for a slow adaptation of the glass to gain its equilibrium with the new environment.

At moderate levels of relative humidity, acid pollutants accelerate the degradation of glass, notably in display cases with a high concentration of acetic acid (released from showcase materials or from other objects) where heavy damage has been reported on sensitive glasses even after exposure periods of a few months only.

The attack of humidity and outdoor acid pollutants is an even more important issue for stained glass windows that are exposed in the cathedrals, in their original architectural setting. For these masterpieces, protective glazings are installed as a preventive conservation measure: the original window is moved a few centimetres towards the interior of the church and installed in a new frame, whereas a new transparent and stable glass is placed in the original setting as a weather shield. For these installations the ventilation of the interspace has to be optimised to allow sufficient air circulation and to reduce condensation effects on the original glass window.

Special care has to be taken for waterlogged glasses: fragments excavated from a wet soil environment are normally heavily degraded, exhibiting thick laminated layers (see Figures 7–9). These fragments may have to be temporarily stored in water and then adapted slowly to museum conditions by a special drying process: the water is replaced by water–ethanol mixtures and exchanged gradually with organic solvents that allow consolidation with an organic polymer to stabilise the fragments, before regular storage is possible.

In general, it must be stressed that glass is considered as a relatively stable material, although various groups of objects do have a special need for preventive conservation. Although a large part of a particular collection might be easy to keep, inappropriate environmental conditions may cause severe damage for certain sensitive glasses, such as medieval glass types or water-logged glasses. In order to avoid iridescence or micro-cracking, the relative humidity during storage and display has to take into account the degree of degradation of the objects, their composition and provenance, and possible damage due to previous treatments.

3   DETERIORATION AND CONSERVATION OF CERAMICS

3.1   History and Technology

Manufacturing of ceramic objects, both for daily use and to express artistic or religious merit, represents an ancient craft, dating back to at least 10000 BC, but still influencing modern technology. Ceramic artefacts are characteristic for the artistic expression of a certain culture and thus have been used by archaeologists as indicators for a social unit and for trade between different units. Ceramics testify cultural activities by providing traces of mankind back to before the Neolithic age and thus being of special importance for civilisations from which no written documents are available.

Although the basic technology for the production of ceramics is rather simple, it leads to a large variety of shapes and objects with characteristic features: by adding the right amount of water, clay is obtained as a soft plastic mix that can be formed easily by hand, *e.g.* to produce simple pots. The walls of a vessel may also be built up by a series of coils, winding round and round a flattened pad. On the other hand, flat sheets of rolled clay can be joined to obtain more rectangular pieces. Figurines, and other complex hollow forms, are produced by pouring liquid clay into an absorbent mould, made from low-fired, unglazed clay or wood. As the water is absorbed by the mould, a thick layer of clay precipitates on its inner surface. After the remaining liquid slip has been poured away, the mould can be opened to remove the object. Different components can be cast separately before joining together with a slurry of liquid clay. The invention of a rotating turntable or wheel has facilitated the mass production of vessels. "Throwing" a pot involves the centering of a ball of clay on the rotating table and forming of the vessel walls with both hands.

The characteristic variety of ceramics is not only created by the different possibilities of shaping clay, but also through variation of raw materials, decoration and firing technology.

3.2   Raw Materials

The Earth's crust is composed of clay minerals and accessory minerals such as quartz, feldspar and calcite. The clay mineral portion consists of particles essentially smaller than 2 μm, whereas the accessory minerals may range from $2\,mm$ down to $2\,\mu m$. Clay minerals exhibit a very large surface area per unit weight, which explains their high attraction for water and other substances.

Whereas soil scientists would classify clay by considering the particle size, a chemist describes it as part of the family of silicates and more precisely as hydrated silicates of aluminium. However, the classification of minerals with a stoichiometric formula does not reflect the complexity of their structure. In order to explain the properties of clay minerals in an adequate way, the coordination of silicon has to be described first, before looking at a long-range order in the crystals.

The basic structural unit of silicates is the SiO_4 tetrahedron that may be bonded to other cations only (*e.g.* as orthosilicates). However, SiO_4 units may also share one or more oxygen atoms at the corner of the tetrahedron to form rings or chains. Most of the clay minerals consist of infinite sheets of SiO_4 tetrahedra, which explains their plasticity and absorption capacity for water between the two-dimensional network. In contrast to this, a three-dimensional network of SiO_4 units leads to stable and hard materials, with no attraction to water. Crystalline quartz is the simplest example for such a compound, building the bridge to glass (see above).

The SiO_4 tetrahedon can be regarded as an ion charged as SiO_4^{4-}. Tetrahedra are linked by sharing corners through Si—O—Si bridges. Clay, based on a two-dimensional network connects tetrahedra that share three corners, thus reflecting the general formula $[(SiO_5)_n]^{2n-}$. The stacked sheets are held together by cations and hydroxyl groups, in order to obtain electrical neutrality in the mineral. One point to note is that aluminium can be incorporated in the mineral for partly replacing silicon in the plate-like laminae, but it can also act as interstitial cation, which further complicates the general formula. As a relatively simple example, kaolinite should be mentioned, with the chemical formula $Al_2(SiO_5)(OH)_4$ (Al is bonded here only as cation between the flakes).

When using the term "clay" as raw material for ceramics, this usually refers to a mixture of different clay minerals, with minor amounts of other non-clay minerals, such as quartz. In terms of their future potential use, these clays are classified, *e.g.* as "white-burning clays" (used for whitewares), stoneware clays (containing fluxes) or brick clays (containing iron oxide). Fillers and fluxes can be added to improve plasticity, to reduce shrinkage during drying or to reduce the firing temperature.

The water content of the clay has to be gradually reduced before firing to avoid fracture. Evaporation at room temperature leads to a dry clay consisting of a porous open structure. This is the best moment to clean the pottery, to join pieces together or, to decorate the surface. By heating up dried clay to temperatures above $800°C$, one obtains ceramics or pottery.

Figure 12 *Fragment of a Neolithic ceramic vessel (Romania, around 5000 BC), an overview*

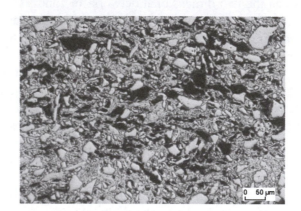

Figure 13 *Fragments of Neolithic ceramic vessel (Romania, around 5000 BC), a cross section; SEM: the coarse structure with partial fusion of the grains is visible*

3.3 Firing Process

The firing process is the key step in the manufacturing process: the porosity of the material is reduced and particles are connected by partial fusion and/or sintering of amorphous constituents (Figures 12 and 13). Firing strengthens the structure and determines the major characteristics of the ceramics, such as the colour and the resistance against environmental attack.

The basic principle in turning a paste with high plasticity into a hard and durable material is connected with dehydration: at room temperature most of the absorbed water is already lost from the surface of the crystals. At a few hundred degrees, the water trapped between the layers is removed and later, structural OH groups are attacked. By thermal analysis it could be shown that this process depends not only on the crystal structure of the mineral but also on the grain size. Various dehydration reactions lead to the formation of high temperature alumino–silicate phases, such as mullite ($3Al_2O_3 \cdot 2SiO_2$) or crystobalite (a form of silica, which is stable above 1470°C). To simplify the composition of ceramics for classification, the different phases present are summarised in triangular diagrams (*e.g.* CaO—Al_2O_3—SiO_2 for earthenware).

Ceramics can be classified by considering the firing temperature and the resulting porosity: thus high-fired stoneware (produced at above 1000°C) has porosities less than 2% and low-fired earthenware (firing between 600 and 900°C) with far more than 10% porosity are at the upper and lower ends of the scale. Porcelain (defined as white and translucent ceramic, fired up to 1400°C) can exhibit an extremely low porosity, whereas terracotta or raku (both fired below 1000°C) would be examples of high porosity.

Various types of kilns evolved during the history of ceramic production: fired brick walls were constructed for field kilns in contrast to periodic kilns, serving only for a limited production campaign; round or tunnel kilns provide different features concerning the duration of the firing cycle or the temperature uniformity. A kiln with an oxygen-deficient atmosphere produces harder ceramics (and a different colour) than a kiln with an oxidising atmosphere.

3.4 Colouration and Glazes

The choice of raw materials (and additives) as well as the firing technology influences the colour of the final product, resulting in cream, yellow, grey, red or brown bodies. Additional colouration of the body can be achieved by coatings, including sprinkling of sand or salts, through mineral slurries or coloured glazes, which are solidified in a second firing process.

Glossy surfaces are obtained by dipping or painting the ceramic body with a slip, which is actually a diluted clay mixture with a similar composition to the body clay. Ancient Greek and Roman pottery are well-known examples for this technique.

Glazes are basically a thin layer of glass (around 0.2–0.4 mm thickness) fired onto a ceramic body. They can be applied by mixing directly the raw ingredients in water or via a frit which is a suspension of ground glass powder. In "lead glazes" the major flux is a lead compound, whereas "tin glazes" refer to lead-containing glazes, opacified with tin oxide. "Alkaline glazes" are very similar to medieval window glass (with Na_2O or K_2O, CaO and SiO_2 as the main components) (see earlier).

Glazes are not only used to decorate art objects or to create a dense surface for storage vessels. Glazes are also popular for colouring of architectural elements and facades (Figure 14). As an example, a glaze used to obtain a brown–green glossy surface on a brick would contain (all numbers in wt%): 48 PbO,

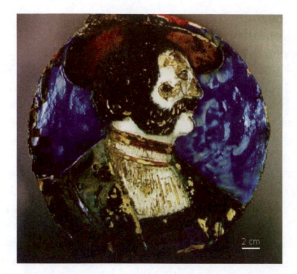

Figure 14 *Glazed terracotta, an architectural element from a 16th century building in Northern Germany*

40 SiO_2, 1 K_2O, 3 Al_2O_3, 1 Fe_2O_3, 3 CaO and 2 CuO (two others). Thus, the main components and even the colouring agents used are again very similar to glass. Various oxides of Fe, Cu, Co, Mn and Sb are considered traditional colourants for antique ceramic glazes, whereas a larger variety of coloured glazes have been available since the 18th century. Fine decorations can be achieved by painting the colourants directly on the ware before glazing, as "in-glaze colour" or "overglaze", when applied on a glazed ground.

For the stability of the composite material, it is important that the body and the glaze have similar expansion coefficients, to avoid cracking and crazing. The bond between the glaze and the underlying body is often weak, for low-fired glaze. In contrast, high-fired glazes can stabilise a friable ceramic support. Firing of glazes onto ceramic bodies implies a sequence of chemical and physical reactions, leading to the formation of an interface, which represents a transition between the dense surface and the porous substrate (Figure 15). The adhesion is ideal if the surface tension of the applied glaze is low enough to enter the pores to create the connection. On the other hand, components from the ceramic substrate may diffuse into the glaze during firing. The thickness of this transition layer may reach up to 50 μm and depends on the time and temperature during firing.

3.5 Dating and Provenance

The typology of pottery styles has been the most reliable dating method for archaeological sites for centuries. Through comparison of characteristic pottery

Figure 15 *Glaze on ceramic, a cross section; SEM: the glaze is well linked with the ceramic support (within the dense glaze cover, a substructure, including bubbles caused by the production process, is visible)*

types, considering the evolution of form and decoration, a sequence in time and local stratigraphy of sites could be established. For the Roman period, the typology is so well defined that dating within a range of 25 years is possible. Trade with pottery from sites with known chronology helped to characterise undated cultures, but also to establish economic and social factors within communities.

A first characterisation of pottery by visual means would include the structure of the fabric (fine or coarse), its colour (red or grey, different layers?) by looking at the core of a freshly broken piece and by looking at the surface (polished or colour coated?). A stylistic approach would note indications of the form, *e.g.* of the rim, of the pedestal base and the handles. However, identical figure types can be produced in different places or at least from raw materials imported from different sources.

Therefore, chemical analysis is the key to provenance studies, especially for ceramics with identical or similar figure types, for pottery with an unknown origin or when trade routes are to be investigated.

The provenance of ceramics can be determined by using major and minor element analysis, obtained mainly by atomic absorption spectrometry (AAS) or X-ray fluorescence (XRF). Even more common is the interpretation of trace element patterns (concerning elements present at less than 0.1 wt%), for which neutron activation analysis (NAA) and inductively coupled plasma spectrometry (ICP–AES or ICP–MS) are the most commonly used analytical techniques.

Different clay sources can be distinguished by absolute concentrations of trace elements, or through the degree of correlation between pairs of elements (*e.g.* Fe and Sc). As an example, the most famous Roman clayware, characterised by a dark red glossy surface, called *terra sigillata*, can be distinguished from fine pottery produced in the Rhineland by looking at the content of As, Zn, Rb and Sr. Cluster analysis of chemical compositions of sherds from one production site may even allow discrimination of wares produced in adjacent kilns (*e.g.* based on the Cr, Mg and Ca content).

Apart from chemical analysis, thermoluminescence (TL), isotope analysis, Carbon 14 (^{14}C) dating and several methods using thin sections should be mentioned here as further techniques applied to ceramics.

Analytical studies of archaeological ceramics are still in progress to obtain improvements, *e.g.* to include possible changes in composition during firing of clay or post-depositional changes in ceramics. The outcome of provenance work depends strongly on the collaboration between archaeologists and analytical chemists, since the reliability of data is based on the selection of samples for analysis and on the profound description of the historical context.

3.6 Deterioration

In general, ceramics are rated as durable materials, which persist even thousands of years of burial in the soil. Nevertheless, for specific objects, ageing may produce changes in colour, accumulation of dirt or under more severe exposure conditions, crumbling and even disintegration.

The rate of degradation is a function of composition, pore structure, manufacturing procedure, structural design, surface finish and possible damage deriving from the time of daily use. Deterioration of ceramics results from several mechanisms, including freezing and thawing, salt crystallisation, chemical attack by water and other substances, various expansion reactions or due to a mismatch of various components within one object.

The porosity of archaeological ceramic is the key factor for post-depositional deterioration, as it allows soil solutions to penetrate and attack the ceramic body. For one specific case, it was shown that a low pH of the burial environment reduced the hardness of pottery. Furthermore, rehydration is postulated as a degradation mechanism, reversing the loss of water during firing and leading to disintegration. Low-fired ceramics (below 700°C) may undergo mineralogical change, leading to the formation of, for example, carbonates, hydrosilicates or gypsum. Water may also cause damage by removing soluble phases from the body, such as calcite ($CaCO_3$), occurring in pottery originally tempered with limestone. Calcareous ceramics are sensitive to deterioration, even when exhibiting low porosity, which is due to the solubility of lime-rich silicates and the low firing temperatures of this type of ceramics. Under extreme circumstances, weathering may lead to a mass loss of up to 20% of an object. On the other hand, penetration with soil solutions may lead to precipitation of

Figure 16 *Salt efflorescence on ceramic body (shown by optical microscopy of the surface)*

salts in the pores (Figure 16). More severe changes are to be expected with seawater, being an even more aggressive chemical environment.

For architectural elements, the access to rainwater or rising damp is decisive for degradation. Here, salt crystallisation is an even more damaging factor than for archaeological ceramics. Water-soluble salts can enter the body, and then recrystallise during drier periods. The expansion in volume of the freshly-formed crystals causes stress to the porous host structure, depending on the temperature and the relative humidity of the environment. Every salt has a critical relative humidity at which crystallisation occurs, although this can be affected by the presence of other salts. Chlorides, nitrates, sulfates and phosphates are the most important soluble salts to consider. The worst scenario is an environment with constantly changing temperature or relative humidity, where salts can precipitate in cycles. Salts are introduced into ceramics at the exposure site, but also during conservation treatments, *e.g.* by using acids for removing stains.

Damage caused by temperature changes can also be connected to frost and thaw cycles, due to the formation of ice in the pores, thus it is more important for architectural ceramics, exposed to severe weathering conditions.

The degradation of glazes follows similar mechanisms as described for glass corrosion, based on leaching and dissolution reactions, dependent on the pH of the attacking solution. Degradation also leads to similar phenomena, such as micro-cracking or iridescence effects. In the case of glazed ceramics, the adhesion of the glaze is an important stabilisation factor, although salt precipitation can lead to exfoliation of even large surface areas (Figure 17). Firing of glazes at lower temperatures allows a broader choice of colours, but it makes the surface more susceptible to an environmental impact. A special phenomenon observed on lead glazes is described as blackening: leached lead ions form lead sulfide, through a reaction with sulfides in the soil, resulting from bacterial metabolism.

Finally, it should be mentioned that apart from chemical and physical deterioration processes, damage on ceramics occurs very often simply by fracturing, during handling in domestic use or due to disasters, such as floods or fires.

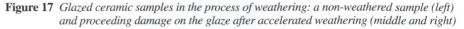

Figure 17 *Glazed ceramic samples in the process of weathering: a non-weathered sample (left)
and proceeding damage on the glaze after accelerated weathering (middle and right)*

Fracture occurs at the weakest joint of an object, which connects this type of damage with the production process. The resistance to mechanical wear is a function of the surface hardness; once scratches have occurred, they represent prominent points for further attack.

3.7   Restoration and Conservation

Among all the materials used to create decorative arts, ceramics are the most stable ones. Nevertheless, the damage phenomena described above occur on ceramic objects; whereas some collections or groups of finds are more affected, others are more stable, depending on the sensitivity of the objects and on the damage potential of the previous exposure or storage conditions. More severe problems occur if ceramics are combined with other materials, such as wood, metals or organic materials within one object. Each of those cases requires special consideration.

Storage conditions for ceramic objects should be optimised to minimise further decay. First of all, fluctuations of temperature and humidity should be avoided to prevent salt crystallisation. Therefore, the general rule is to keep objects in a dry environment, especially if retrieved from a dry exposure site. Objects from damp environments might have to be kept wet, if an immediate treatment is not possible. Removing damaging salts by long-term soaking in water is a common method for future stabilisation of such materials. Air pollutants increase the damage by chemical attack, especially on sensitive glazes and should be therefore reduced in showcases and galleries. Extreme temperature changes and vibrations, causing thermal or mechanical shock, have to be avoided. In contrast, light is an environmental parameter that does not play an important role in ceramics degradation. Monitoring of light is necessary only for very special coloured low-fired ceramics or to prevent degradation of previously applied conservation materials.

Since human intervention has a high damage potential for ceramics, handling and packaging of ceramics deserve special care. Recommendations for lifting and moving objects as well as for the design of containers and transport devices are available.

Cleaning of an object might be necessary to remove encrustations, developed during exposure to soil, such as calcite, gypsum or silicates. Porous pottery might be stained with iron compounds or deposits, resulting from the content of vessels. Difficulties occur, when the "dirt" is soluble and has been carried into the pores of the body.

"Repair" is considered as an early conservation intervention, for ceramics not only as precious artistic objects, but also as objects of daily use, such as household ware. For the conservator today, early repairs might provide a source of valuable information and might be kept to document the history of the piece before burial. Other, more recent and aesthetically disturbing bonding, filling or retouching materials might have to be removed, thus representing serious restoration problems. Removal of previous adhesives or coatings is necessary if shrinking or yellowing of the polymer is affecting the artefact. Mechanical tools, *e.g.* a scalpel, and a wide range of solvents are in use to remove animal glues, bitumen, shellac, cellulose nitrates, silicones, polyvinyl-acetates or epoxies from ceramics. Whereas the cleaning process might be successful for surface coatings, the removal of organic polymers, formerly applied as consolidants and thus penetrating the porous ceramic body, is hardly feasible.

Due to the limited reversibility of consolidation treatments for ceramics, new consolidations are carried out only if the object is seriously endangered, *e.g.* when the ceramic body is crumbling or the surface decoration is flaking off the surface. Materials, such as Paraloid B72 (already mentioned for glass conservation) or silanes and siloxanes (used for stone conservation) can be applied by brush, injection, spray or by immersion. Vacuum impregnation might be useful to achieve deep penetration with the consolidant.

Fragile ceramics often need bonding of broken pieces. The choice of the adhesive has to consider the general properties of a polymer, such as durability and colour match, but also practical considerations, such as the viscosity and the strength of the required bond. For fractured pieces with disturbing gaps, filling with colour-matched polymers is quite common. Fillings can be disguised even better by retouching, if ethically acceptable for the object.

Conservation of ceramics, although dealing with relatively stable materials, is a well-developed discipline. Advanced active and preventive conservation strategies are applied to preserve the extensive collections of ceramics in museums all over the world, covering objects from several thousands of years of ceramic production.

4 CONCLUSIONS

Glass and ceramics cannot be described with a simple stoichiometric formula, although both consist mainly of silicon and oxygen. Both are obtained from very common raw materials, such as sand and clay. Structural changes during firing turns them into solid and rather durable materials. The complex technology behind this simple equation was invented several thousands of years ago and gave artists the opportunity for creativity and personal expression.

Whereas ceramics are polycrystalline materials, consisting of several phases (discrete chemical compounds), glass consists of a three-dimensional network with short-range order. For both, deterioration can occur early due to daily use or much later during exposure in museums, or due to aggressive soil environments for archaeological objects or due to weathering of architectural elements.

Chemical changes on the surface might cause micro-cracks, leading to a dull appearance of glass. Salt crystallisation can endanger the structure of ceramics, which may result in crumbling of the ceramic body. For long-term conservation, a glass might require surface protection, whereas the porous ceramic body might need an in-depth consolidation. Although glass and ceramics, both derived from silica, are so similar in chemical composition, the requirements for conservation can be rather different. For glazed ceramics, representing a combination of both materials, a practicable compromise has to be found for each specific case.

REFERENCES AND FURTHER READING

D.R. Brothwell and A.M. Pollard, *Handbook of Archaeological Sciences*, Wiley, New York, 2001.

S. Buys and V. Oakley, *The Conservation and Restoration of Ceramics*, Butterworth-Heinemann, Oxford, 1993.

D.E. Clark and B.K. Zoitos (eds), *Corrosion of Glass, Ceramics and Ceramic Superconductors*, Noyes Publication, Park Ridge, NJ, 1992.

S. Davison, *Conservation and Restoration of Glass*, Butterworth-Heinemann, Oxford, 2003.

H. Scholze, *Glas – Natur, Struktur, Eigenschaften*, Springer, Berlin, 1988.

N.H. Tennent, *The Conservation of Glass and Ceramics*, James & James, London, 1999.

H. Wedepohl, *Glas in Antike und Mittelalter*, E. Schweizerbart'sche Verlagsbuchhandlung, Stuttgart, 2003.

A. Wolff, *Restaurierung und Konservierung historischer Glasmalereien*, Verlag Philipp von Zabern, Mainz, 2000 (to be translated into English).

CHAPTER 8

Plastics

YVONNE SHASHOUA

Department of Conservation, National Museum of Denmark, PO Box 260 Brede, DK-2800 Kongens Lyngby, Denmark

1 PLASTICS IN HERITAGE COLLECTIONS

Synthetic plastic materials have had a significant influence on industrial, domestic and cultural aspects of modern life throughout the 20th, and into the 21st, centuries. They represent advances in technology, illustrated by the dramatic growth in information storage media available in the last 20 years, development of space suits, credit cards and food containers that can be taken directly from the freezer to microwave oven and then to dinner table without failing. It is interesting that until the 1940s, it was not possible to drink hot coffee from a plastic cup without softening it – an activity we consider commonplace today.

The development of plastics also reflects economic history. Restrictions on imported latex, wool, silk and other natural materials to Europe during the Second World War resulted in the rapid development of alternative synthetic plastics. Table 1 shows that between 1935 and 1945, many new polymers were introduced including polyethylene, polyamides, poly(methyl methacrylate), polyurethanes, poly(vinyl chloride) (PVC), silicones, epoxies, polytetrafluoroethylene and polystyrene. Polyethylene was incorporated into radar systems while PVC replaced the limited stocks of natural rubber as cable insulation.

Public attitudes towards plastics have also changed. When the first man-made plastics formulation, cellulose nitrate, was exhibited at the Great International Exhibition in England in 1862 by Alexander Parkes, it was designed to imitate luxury materials, such as tortoiseshell and ivory, which were, at that time, in increasing demand and diminishing supply. However, this image of plastics as highly-valued luxury goods faded when the colourful, post-Second World

Table 1 *First commercial availability of plastics*

Date	Plastics type	Uses	Comments
1862	Semi-synthetics Cellulose nitrate	Synthetic ivory, tortoiseshell, film base	Rayon, Celanese, Tricel
1894	Cellulose acetate	Doping for aircraft wings, 'safety' film, synthetic textiles	Rayon, Celanese, Tricel
1902	Casein-formaldehyde	Buttons, hot-stamped products	Erenoid, Galalith, Ameroid
1909	Synthetics Phenolformaldehyde	Radio cases, cables, insulators, heated hair rollers, fine boxes, photo	Bakelite, Catalin 'The material of 1000 uses' Only made in dark colours
1926	Urea and thiourea-formaldehyde	Kitchenware	Available in pale colours
1924	Melamine-formaldehyde	Formica laminated surfaces	Scratch-resistant and hard
1935	Poly(vinyl chloride)	Cable insulation, Barbie dolls, toys	Replaced rubber in Second World War. Needs lots of additives
1937	Poly(methyl methacrylate)	Aircraft windows, glazing, lighting	Plexiglas, Perspex
1938	Polystyrene	Heat insulation, plastic cups	Flamingo foam
1939	Nylon	Parachutes, synthetic silk, clothing	Very tough, withstands hot water
1942	Polyethylene	Tupperware	Melts at 130°C
1947	Epoxy	Adhesives, sculptures	Yellows quickly
1950	Polyester	Fleeces, cola bottles	Polyethylene terephthalate
1956	Polypropylene	Car bumpers, bottle crates	Stiffer than polyethylene
1967	Polycarbonate	CDs, greenhouse roofs	Tough

War designs were marketed in large numbers as housewares. They developed a long-lasting image as low value, low quality and ephemeral pieces. It has even been said that dreaming about plastics suggests that one is fake and artificial!

The 1980s saw a change in perception of plastics from disposable materials to fashionable, highly collectable pieces with historical and technological significance. Expansion in the number of processing and fabrication techniques has allowed modern plastics to be manipulated in thin film, bulk and foam forms, and to be combined with fibres, metals and wood. Today there are approximately 50 different basic types of plastics, included in 60,000 different plastics formulations; those based on polyolefins and PVC have highest consumption worldwide. Six new plastics formulations are sent for evaluation and approval to major testing laboratories each week.

Museum objects are rarely collected for their material type but because of their origin, function, design, rarity, cultural or historical significance; plastic objects are collected for the same reasons. In this way, museums act, often unintentionally, as storage depots for both early plastic materials and the most recent experimental formulations. Some early plastics are no longer used on a commercial scale; this may be due either to concern about their flammability, toxicity or because their performance is considered inferior by today's standards. An example of this is cellulose nitrate that is highly flammable, and therefore no longer permitted for use in public buildings or transport in Europe. As a result, it is manufactured by only a few companies today. However, cellulose nitrate can be found in the form of adhesive, spectacle frames, jewellery and table tennis balls in museum collections.

Today, almost all international museums and galleries possess collections which contain plastics. Plastics may be identified within building materials, defence equipment, ethnography, furniture, housewares, information technology, medical and sports equipment, modern art, photography and toys. Many combine metals, textiles and wood with plastics in their construction. In addition to the objects themselves, many of the materials used to store, transport and display them are also plastics. While museums continue their policy to collect objects that reflect both everyday life and historical events, the proportion of plastics in museums will increase.

Once plastic objects are registered in museum collections, the institution becomes responsible for their long-term preservation, until the end of their useful lifetime; that point is reached, arguably, when the object ceases to have a recognisable form or meaning. Most plastics have rather a short lifetime compared with those of traditional 'craft' materials found in museums, such as ceramics and stone. The definition of useful lifetime as applied to plastics in museum objects is rather different to that defined by the plastics industry. In 1954, Quackenbos, an industrial chemist, defined the life of a plasticised PVC film as the period taken to lose 10% of its original weight. After that period, the material is considered to have changed so much in character that it has failed. Calculations based on Quackenbos' research indicate that, depending on formulation, lifetimes for PVC films range from 3 months to 1000 years at 25°C. It is a more complex matter to estimate the life expectancy of a plastic object in a museum. We must first establish how much deterioration is acceptable before an object shows a reduction in quality. While yellowing and other changes in appearance are recognised as normal manifestations of deterioration for objects constructed from natural materials and are usually left untreated, the same changes in plastic objects are usually deemed unacceptable.

Since the properties and degradation reactions of plastics are usually influenced by their formulations and manufacturing processes, this chapter will first discuss those areas. Four plastics have been identified as being more vulnerable to degradation than others in museum collections; cellulose nitrate, cellulose acetate, plasticised PVC and polyurethane foam. The most frequently seen degradation reactions and the possibilities for conserving these plastics will be presented in this chapter.

2   THE CHEMISTRY AND PHYSICS OF PLASTICS

According to the Penguin English Dictionary, the noun 'plastic' is defined as 'any of the numerous synthetic organic polymers that, while soft, can be moulded, cast, *etc.* into shapes and then set to have a rigid or slightly elastic form'. In practice, although the polymer component of a plastic is always the greatest by weight, it is usually modified by adding other materials to improve inherent stability, longevity, physical and chemical characteristics.

   Polymers are used as the base material to make plastics. Polymers are large molecules often described as chains, with a characteristic 'repeat' chemical unit acting as the links. The repeat unit is usually a small molecule known as a monomer; typically there are between 1000 and 10,000 monomers in a polymer chain. The number of repeat units can be used to describe the chain length of the polymer, but the use of molecular weight is more common. Monomers may contain carbon, hydrogen, oxygen, nitrogen, silicon, chlorine, fluorine, phosphorous and sulfur. The chemical process of joining the monomers into a high molecular weight polymer is called polymerisation. Since polymerisation usually produces chains of varying lengths, the molecular weight is calculated as an average weight per chain known as weight-average molecular weight (M_w).

   Polymers are often named after the monomer rather than after the repeating unit in the structure; thus the polyethylene molecule, prepared from the monomer ethylene, consists of a long chain of repeating *methylene* —(CH_2)— groups (Formula 1). Polymers prepared from more than one species of monomer are called copolymers, *e.g.* butadiene-styrene polymers.

Formula 1. Polyethylene consists of a long chain of repeating methylene groups.

$$\sim CH_2{-}CH_2{-}CH_2{-}CH_2{-}CH_2{-}CH_2{-}CH_2{-}CH_2{-}CH_2 \sim \qquad (1)$$

2.1   Preparation of Polymers

The first polymers were developed in 1862, known as semi-synthetics and formed a technological bridge between natural (those produced by trees, plants and insects) and fully synthetic polymers. Semi-synthetic plastics were made by treating a natural material chemically to modify its properties, usually with the aim of producing a mouldable product. In 1909, the first fully synthetic polymer was produced by reacting two chemicals (monomers) together.

Preparation of semi-synthetic polymers. Cellulose plastics, particularly cellulose nitrate and acetates, were the most commercially-important semi-synthetics, and have been used to prepare photographic films, textile fibres and lacquers.

They are made by esterifying cellulose, in the form of either cotton linters or wood pulp. Cotton linters, the short fibres removed from cotton seeds, are purified to remove impurities and coloured entities by heating at 130–180°C under pressure in a 2–5% aqueous solution of sodium hydroxide. The resulting cellulose has a molecular weight in the range 100,000–500,000. Non-cellulosic material is removed from wood pulp, most frequently via the sulfite process, which involves treatment with a solution of calcium bisulfite and sulfur dioxide. Each of the thousands of repeat units of pure cellulose has an empirical formula $C_6H_{10}O_5$ and, despite the abundance of hydroxyl groups, is not water-soluble due to its high crystallinity and extensive hydrogen bonding.

Reaction Scheme 1 shows that during preparation of cellulose plastics, the hydroxyl groups on each cellulose monomer ring are replaced by other substituent groups. To prepare cellulose nitrate, pre-dried cotton linters are treated with concentrated nitric and sulfuric acids. Water tends to slow the reaction, so its presence is limited. Sulfuric acid catalyses the reaction, so it can occur without heating and under ambient conditions. The product is washed with water to remove the residual acids; if residues of sulfuric acid are allowed to remain, an explosive reaction can occur. Remaining water in the cellulose nitrate is displaced by alcohol.

The likelihood of a particular hydroxyl group being replaced is largely determined by its position in the monomer molecule. Substitution of all three hydroxyl groups on repeat units of the cellulose molecule results in the creation of explosive cellulose trinitrate, containing 14.4% nitrogen as shown in Reaction Scheme 1. Cellulose dinitrates containing between 11% and 13% nitrogen (an average degree of substitution between 1.9 and 2.7 hydroxyl groups) are useful for plastics, photographic film bases and lacquers. New cellulose nitrate is a colourless polymer, which is brittle and tough, but flexible when plasticised and cast as thin films and sheets. Cellulose nitrate is highly soluble in polar solvents, such as acetone, ethanol and butyl acetate. It softens at approximately 80–90°C and is thermoplastic, flows at 150°C, and ignites in air at 160°C. The high flammability of cellulose nitrate was a highly undesirable property in its applications as coating, doping (tightening and wind-proofing) treatment for the canvas-covered wings of aircraft used in the First World War and in cinematographic films; it resulted in the development of other cellulose esters, most important of which was cellulose acetate at the start of the 20th century.

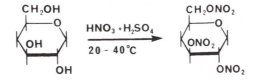

Reaction Scheme 1 *Preparation of cellulose nitrate involves replacing the hydroxyl groups on each of the many monomer units of cellulose*

The acetylation of cellulose to cellulose acetate involves three stages, pre-treatment of the cellulose, acetylation and hydrolysis. The aim of pre-treatment is to purify the raw material and open the fibrous cellulosic structure to facilitate substitution. Exposure to glacial acetic acid is the usual pre-treatment. The pre-pared cellulose is mixed with a slurry of glacial acetic acid, acetic anhydride and sulfuric acid for 5–6 h, depending on the desired degree of substitution. Since the acetylation reaction is exothermic, the acids must be pre-cooled to limit the tem-perature rise. As the reaction proceeds, the fibres of cellulose, insoluble in glacial acetic acid, form soluble cellulose acetate. Just as for cellulose nitrate, various degrees of substitution of hydroxyl groups are possible. Cellulose diacetate com-prises a degree of substitution of 2.2–2.3 or 36–38% acetyl ($-COCH_3$) and is suitable for moulding, while cellulose triacetate has a degree of substitution 2.8–3.0 (42–44%) acetyl and is used for photographic film and fibres.

Preparation of synthetic polymers. The polymerisation process (chemical joining of monomers) generally occurs by means of one of the three major mechanisms, namely, addition polymerisation, condensation polymerisation and rearrangement polymerisation.

In addition polymerisation, monomers are converted into polymers without the formation of side products. Polymers formed by addition polymerisation include polyethylene, polypropylene, poly(methyl methacrylate), polystyrene and PVC. Reaction Scheme 2 shows that for addition polymerisation to take place, the monomer must contain at least one double bond. Double bonds are broken in the first stage of the process, known as initiation, and the released valences join to other monomer molecules. A catalyst may be necessary to initiate or hasten the process (such materials include benzoyl peroxide and azodi-isobutyronitrile) which is encouraged to decompose into reactive free radicals by exposure to light or heat. The catalyst's radicals react with monomer molecules by addition, generating even more radicals which further react with monomer molecules. In the second stage or propagation, monomers continue to add together, forming increasingly longer chains. During the final stage, known as termination, all the monomer molecules have reacted and the reaction ceases. The reaction may also be terminated rapidly by quenching with water. The degree of polymerisation or number of repeat units in the molecular chain, n, ranges between 500 and 1500; this corresponds to a theoretical molecular weight average of 100,000–200,000. In practice, all batches con-tain molecules with a range of chain length.

$$nCH_2 = CHCl \longrightarrow (-CH_2 - CHCl-)n$$

vinyl chloride   poly(vinyl chloride)

Reaction Scheme 2 *Poly(vinyl chloride) is formed by addition polymerisation of vinyl chloride*

The process used to produce more than 80% of commercial PVC polymers is suspension polymerisation. An aqueous suspension of vinyl chloride monomer is agitated vigorously in a pressurised vessel together with colloids (detergents) to hold monomers in suspension, and buffers to control pH. The resulting PVC particles are roughly spherical and range from 50–250 μm in diameter.

Commercial PVC is essentially an amorphous material, although a small amount of crystallinity is present (about 5% as measured by X-ray diffraction methods) and is attributed to the fact that that the bulky chlorine atoms do not align and pack readily. Despite this low percentage, crystallinity greatly influences the properties of PVC in solution and solid phases.

PVC is thermoplastic, so it softens on warming. The presence of the chlorine atoms in the structure increases the attraction between chains due to C—Cl dipolar interactions, imparting great hardness and stiffness in the polymer. The glass transition temperature of commercial grade PVC is approximately 80–84°C and the melting point around 212°C. Maximum service temperature for commercial compounds is between 65°C and 80°C. The polarity introduced by the presence of chlorine means that PVC is soluble in polar solvents; examples are tetrahydrofuran and cyclohexanone. The high concentration of chlorine in the polymer (56.8% by weight) imparts flame-retardant properties, a property which has been utilised in electrical insulation cables and housings.

During polymerisation by the condensation technique, a reactive chemical group on one monomer reacts randomly with another group on a second monomer molecule, with the formation of a small molecule as the monomers join together to form a chain. The small molecule is usually water or alcohol that splits off during polymerisation. Since the reaction is only dependent on monomer molecules coming in contact with others, a significant increase in average molecular weight will occur with reaction time. Polymers formed by condensation polymerisation include polyesters, nylons and formaldehyde polymers.

Rearrangement polymerisation may be considered to follow a reaction type intermediate between those of addition and condensation. In common with addition polymerisation, there is no splitting off of small molecules, but the kinetics are otherwise similar to condensation processes. The preparation of polyurethane polymers occurs by rearrangement polymerisation. Polyurethane polymers are products of a polyol, based either on a polyester or polyether, with several alcohol groups (—OH), a di-or poly-isocyanate with several cyanate groups (—N═C═O) and a chain extender. The chain extender reacts with the polyol's alcohol groups, initiating an imbalance in negative and positive charges throughout the molecule, which after reaction with the isocyanate, results in the formation of a urethane group (—NHCOO—) as shown in Reaction Scheme 3.

$$\underset{polyol}{HOROH} + \underset{}{OCNR_1NCO} + \underset{}{HOROH} + \underset{polyisocyanate}{OCNR_1NCO}$$

$$\downarrow$$

$$\underset{polyurethane}{-OROOCNHR_1NHCOOROOCNHR_1NCHO-}$$

Reaction Scheme 3 *Preparation of polyurethane from polyol and polyisocyanate*

To produce polyurethane foam, water is added to the polyol and isocyanate starting materials. The water molecules react with the cyanate groups to form amine groups ($-NH_2$) and carbon dioxide gas. The amines continue to react with isocyanate groups to form urea linkages ($-HNCONH-$) between the chains, instead of the urethane groups created in the absence of water. As polymerisation progresses and molecular weight increases, carbon dioxide gas becomes trapped in the increasingly viscous liquid polymer. The trapped bubbles comprise cells in the polyurethane foam.

The physical properties of the final polyurethane are determined by those of the raw materials. The physical form in which polyurethanes are produced, whether as fibres or expanded foams, depends on the chemical formulae of both the isocyanate and the alcohol or components; these control the molecular weights and extent of cross-linking in the resultant polymer. Polyglycols, such as a polyethylene glycol with a molecular weight of around 2000, result in a higher molecular weight polyurethane than if a diol was used. Polyurethanes are noted for their resistance to most organic solvents and high tensile strength. Due to the toxicity of isocyanate vapour, one of the raw materials, good ventilation and safety equipment are necessary when handling these materials.

Most polymerisation reactions do not progress to completion, resulting in the residues of starting materials in the plastic. Approximately 1–3% of residual monomer is found in acrylics, PVC, polystyrene, polycarbonates, polyesters, polyurethanes and formaldehyde polymers immediately after production. Monomers with boiling points lower than ambient are likely to have evaporated before the final product is used, while those with higher boiling points, including styrene monomer and terephthalates used in saturated polyesters, off-gas slowly from the plastic formulation and can often be detected by odour.

2.2 Additives

The chemical and physical properties of polymers can be changed considerably by incorporating additives soon after manufacture. The type and quantity of additives allow many different products to be produced from the same polymer.

Table 2 *Typical applications for PVC formulations
plasticised with di(2-ethylhexyl) phthalate (DEHP)*

DEHP (% by weight)*	Application
16.7	Vinyl flooring
23.1	Upholstery cover
28.6	Document folder
33.3	Garden water hose
37.5	Electrical cable sheath
44.4	Shoe sole
50.0	Rubber (wellington) boots

*Based on PVC and DEHP.

For example, raw PVC polymer is a brittle, inflexible material with rather limited commercial possibilities. Attempts to process PVC using heat and pressure result in severe degradation of the polymer. Compounding PVC involves incorporating sufficient additives with the raw polymer to produce a homogeneous mixture suitable for processing and with the required final properties for the lowest possible price. Table 2 shows that PVC can be used to produce rigid guttering, plastic grass, soft toys, electric cable insulation, photograph pockets and shoe soles simply by varying the amount of plasticiser in the formulations between 16% and 50% by weight.

During processing, PVC is milled together with liquid plasticiser, so that the latter physically attaches itself to the surfaces of the polymer particles, separating them from each other and allowing them to flow over each other, increasing flexibility on a macro-scale. Of the one million tonnes of plasticisers used annually in Europe, approximately 90% comprise phthalate esters; the largest single product used as a general purpose plasticiser since the 1950s is di(2-ethylhexyl) phthalate (DEHP). The effects of DEHP on health, particularly that of children, has been of concern since the 1980s, and toys and accessories containing DEHP intended for children younger than three years are not allowed to be sold in Europe.

Polymer additives or modifiers may be generally grouped into those which alter physical properties, those which change chemical properties and those which improve the appearance of the finished plastic product. Additives, which alter physical properties either during processing or in the final plastic object include fillers, plasticisers, lubricants and flow promoters, impact modifiers and foaming agents. Additives, which change chemical properties during processing and in the final plastic product include anti-ageing additives. Additives, which improve appearance include colourants in the form of pigments and dyes.

Table 3 summarises the types and function of the additives that are usually incorporated into polymers. There is some overlap between the properties and actions of many polymer additives. For example, plasticisers may also act as impact modifiers, lubricants and stabilizers. Conversely, a commercial

Table 3 *Types and functions of frequently used additives*

Additives	Major functions	How achieved	Examples
Plasticiser	Soften polymer, reduce Tg	Separates polymer chains from each other	Phthalate esters, aliphatic diesters, epoxidised oils, phosphate esters, polyesters
Lubricant	Prevent adhesion of polymer to processing equipment	Sweats out due to low compatibility with polymer	Calcium stearate, normal and dibasic lead stearate
Impact modifier	Reduce brittleness	Adds bulk to compound	Styrene-acrylonitrile, ethylene-vinyl acetate copolymer
Processing aid	Ensure uniform flow and good surface finish	Migrates to surface to improve flow during manufacture	As for impact modifier
Filler	Opacify compound, increase hardness, reduce cost	Changes refractive index and reflective properties, adds bulk to compound	Calcium carbonate, magnesium carbonate, barium sulfate, carbon or glass fibres
Stabilizer or anti-ageing	Minimise or eliminate degrading effects of heat, light or oxygen on polymer	Interrupts free-radical reaction sequence by binding oxygen or degradation product so it is unavailable for further reaction	Tin mercaptides, barium–cadmium salts of fatty acids, lead salts, alkyl benzenes, epoxidised soya bean oil
Colouring agents	Decoration, protection from radiation	Add colour and opacity to surface or to bulk	Carbon particles, dyestuffs

lubricant may act as a plasticiser once incorporated into a formulation. The major requirements for additives in plastics include:

- High stability under processing condition
- High stability when plastic is in use
- Ability to remain in the formulation throughout the intended useful lifetime of the plastic and not to migrate out or evaporate. The intended useful lifetime of most plastics is between one (polyethylene carrier bag) and 30 (unplasticised PVC window frames) years; museums aim to keep them for longer
- Low-toxicity and inert odour or taste, especially in plastics for use with food or medicines
- Low cost.

2.3 Shaping Plastics

Shaping plastics is also known as converting because the aim of the process is to change molten, flowing plastic into a pre-determined shape. Plastics may be converted into films or sheets, solid or hollow three-dimensional forms, fibres and foams. Early plastics, like the natural materials they were developed to mimic, were roughly moulded, hand carved and polished; all resource-demanding activities. As demand for plastics grew, such techniques were replaced by mechanised, automated processes.

Commercially speaking, the most important conversion processes are calendering, extrusion and injection moulding. During calendering, liquid plastic is rolled between heated metal rollers into thin films. Extrusion moulding was first developed in the 1830s. Pellets of thermoplastic plastics are fed into an extrusion mould and heated to melting. The liquid plastic is pressed by a screw arrangement, into a die. Extrusion moulding is suitable for the manufacture of symmetrical shapes of even thicknesses including pipes, fibres and films, where a uniform shape is required. Injection moulding was first developed in the 1930s and varies from extrusion moulding; in the latter process molten plastic is pushed into a closed mould before releasing the cooled, shaped material. Complex, asymmetric shapes, such as plastic water bottles, can be formed by injection moulding.

On cooling, plastic materials tend to contract or shrink considerably more than other materials such as metals, ceramics and glass. For example, a copper pipe will shrink by 0.01% if the temperature is reduced by 10°C. Under the same conditions, a high-density polyethylene pipe would shrink by 0.07%, and polypropylene and hard PVC pipes by 0.04%. In addition, surfaces of plastic materials cool before their cores. Such a situation leads to the initial contraction of plastic materials at surfaces, before significant change in dimension occurs in the bulk. The skins of moulded plastics tend to be stiffer than the bulk, so are more prone to degradation by mechanical action, *e.g.* flexing.

Shaping of foams provides different challenges to shaping of plastics because the foam is growing and expanding while being shaped. The polyurethane polymer ingredients are mixed and poured evenly into a moving trough or Henecke machine. Water and catalysts are then injected into the polymer and the whole is vigorously stirred. As foaming takes place, the mixture forms an even block of foam which, when cool and hard, may be cut to size. If a cylinder of foam is required, the foaming mixture is fed into the bottom of a cylinder and pushed upwards. Pressure applied above the foaming mixture is used to control density. The solidified cylinder may be sliced horizontally into discs.

3 IDENTIFICATION OF PLASTICS

It is important to identify the types of plastics present in objects to understand how, when and why they were produced, and to identify the optimal conservation strategies for them. Methods of identification may be divided into simple tests, based on measuring their appearance, mechanical and chemical properties, and instrumental techniques. Before examining plastic objects, it is helpful to have as much historical information about them as possible. The date of manufacture or, if unavailable, the date of collection, can give information about the type of plastic. If an object was manufactured before around 1905, it is likely to be a semi-synthetic or natural material, rather than a synthetic plastic. Styles are not always helpful in identifying age or material because the same mould can be used for many years and popular styles are often revived.

3.1 Simple tests

Appearance is only a guide to the type of plastic present because various fillers and colouring materials can change the appearance of a polymer considerably. However, Bakelite (phenol–formaldehyde) is always brown, black or red, while urea- and melamine-formaldehydes may have pastel shades. Polyethylene and polypropylene always appear slightly cloudy as thin films, while most other colourless plastics, particularly polyesters and polycarbonate, are crystal clear.

Odour of plastics may assist with identification. Since warming plastics gently increases the concentration of volatile materials, they should be rubbed with a clean cotton cloth just prior to sniffing close to the surface. The smell is usually that caused by volatilising a monomer, plasticiser or degradation product by warming. An odour of vinegar is typical of cellulose acetate while that of mothballs (camphor) suggests cellulose nitrate and a smell similar to that of a new car indicates plasticised PVC. A fishy odour is often produced by warming melamine- or urea formaldehydes.

A rather unusual test is to tap the plastic firmly with a fingernail. If the sound is metallic, the plastic is likely to contain polystyrene. Hardness is another property that can be roughly tested with a fingernail applied to the underside of an object. Plastics which can be marked with a fingernail include polyethylene, polypropylene, plasticised PVC and polyurethane; other types are not affected.

The ability of plastics to float on the surface of a beaker of tap water at 20°C is related to their density at that temperature. The density of water at 20°C is approximately $1 \, g \, cm^{-3}$. If a small sample floats on the surface of the water, it has a density lower or equal to one at the same temperature; if it sinks, it has a density greater than one. Polyethylene, polypropylene and polystyrene float on water while other plastics sink, Bakelite and casein having the highest densities. The flotation test is a rough method to identify plastics since results are dependent on the physical form of the plastic. Foams contain cells filled with air, so their densities will be lower than a solid block of the same type of plastic.

If it is permissible to take a small sample and heat it in a test tube, the pH of the vapours may be determined by placing a piece of moistened

Universal Indicator paper at the mouth of the tube. The results may be interpreted as:

- pH 1–4 (Acidic): cellulose nitrate or acetate, polyester, polyurethane and PVC
- pH 5–7 (Neutral): polyethylene, polystyrene, acrylics, polycarbonate, silicones and epoxies
- pH 8–14 (Alkaline): nylon and formaldehyde plastics.

The flammability of plastics can be a useful tool in their identification, but the additives present have an influence on their ability to burn. A clear test to indicate the presence of chlorine, mainly found in PVC and poly(vinylidene chloride) is the Beilstein test. A clean, copper wire 30–40 cm long, with a cork or other heat-insulating material at one end as a handle, is heated with a Bunsen burner to clean it of residual impurities, heating it until the flame is colourless. The hot, cleaned wire is placed in contact with the plastic to be analysed so that a small piece is melted onto it. The wire is returned to the flame and the colour noted; a green or blue–green flame denotes the presence of chlorine while other colours suggest the plastic to be other than PVC or poly(vinylidene chloride).

3.2 Analytical Techniques

Most simple tests can only indicate the group of plastics to which the object belongs. In order to identify materials more precisely, it is necessary to use instrumental analytical methods. Different techniques provide information about which polymers and which additives are present and their state of degradation, so it is usually necessary to use several in combination. For example, Gas Chromatography–Mass Spectrometry (GC–MS) is a destructive technique that allows identification of the organic components and volatile materials present due to degradation. X-ray Fluorescence (XRF) spectroscopy is an effective, non-destructive surface technique for identifying inorganic fillers, pigments and metal components, but cannot detect polymers.

The most widely available technique for identifying mainly polymer, but also additives in plastics, is Fourier Transform Infrared (FTIR) spectroscopy. Samples are exposed to infrared light (4000–400 wavelengths per centimetre or cm^{-1}) causing chemical bonds to vibrate at specific frequencies, corresponding to particular energies. In the last 5 years, an accessory for FTIR has been developed, which enables non-destructive examination of surfaces and so is ideal for analysis of plastics in museum collections. Attenuated Total Reflection-FTIR (ATR-FTIR) requires samples to be placed on a diamond crystal with a diameter of 2 mm through which the infrared beam is reflected

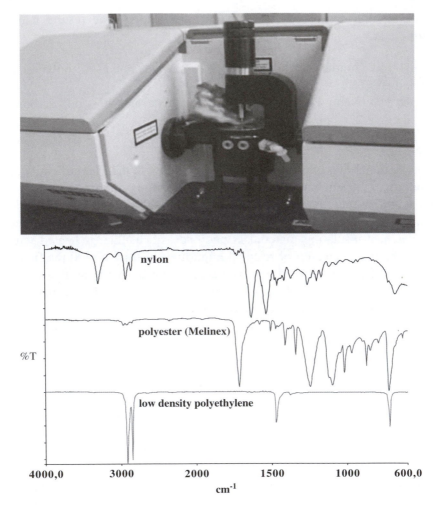

Figure 1 *Attenuated Total Reflection accessory for FTIR spectrometer with diamond crystal through which infrared beam is reflected (upper). Plastic samples must have good contact with the surface of the diamond to obtain well-resolved spectra (lower)*

as shown in Figure 1. The quality of spectra is dependent on intimate contact between the diamond crystal and the sample, so a pressure device is used to achieve this. Diamond has a higher refractive index than most plastics (2.4 and approximately 1.5, respectively), so when the beam comes into contact with the sample, it penetrates only about 2 μm into the material before being totally internally reflected by it into the detector. A detector examines the frequencies absorbed compared with those present in the incident radiation and produces a spectrum or trace which is definitive for all infrared-sensitive chemical bonds present in a particular material or mixture. A spectrum can be obtained 30 s after placing the sample in the instrument.

Unknown plastics may be identified either by comparing spectra with those from reference or known samples; or with reference libraries such as those specific to art, architecture and archaeological materials produced by the Infrared and Raman Users Group (http://www.irug.org); or by using tables to correlate infrared absorption energies with specific chemical bonds. Figure 1 shows that polymers give their own distinctive 'fingerprints', so they can be distinguished from each other. However, FTIR spectra reflect a mixture of all components present in the plastic, including additives and colouring materials, making them complicated to interpret.

4 DEGRADATION OF PLASTICS

A survey of plastics-containing objects in the British Museum and the National Museum of Denmark showed that 1% of objects were actively deteriorating and were in immediate need of conservation, while 12% exhibited deterioration and required cleaning, stabilising and repair. All the 'acute' objects contained cellulose nitrate, cellulose acetate, plasticised PVC or polyurethane foam. Most objects (60%) were defined as being of low conservation priority, that is, they were in a stable condition but needed some treatment such as cleaning. Only just over one quarter required no conservation treatment. In general, deterioration of plastic objects in museums is visible within 5–25 years of collection.

It should be remembered that most museum objects have been used or displayed before they are collected. They have an unknown past, usually, which contributes greatly to the rate and type of deterioration. Instability of the earliest plastics, cellulose nitrate and acetate is expected due to their poorly stabilised formulations and because they are the oldest man-made plastics in museums. However, PVC was first developed in 1926 and is still in use, so its deterioration is rather unexpected.

Causes of deterioration of plastics may be broadly divided into physical and chemical.

Physical causes include:

- *Stress, fatigue and mechanical damage.* These are results of the use of the plastic objects and could comprise frequent bending of a PVC soft toy leading to its failure; or abrasion of the surface of a vinyl record as it is repeatedly pulled out of its sleeve to play, resulting in its inability to produce perfect sound.
- *Migration or loss of additives.* These are consequences of the properties of additives used to formulate plastics; their selection is related to the function, expected lifetime and price of the final product. Camphor, one

of the earliest commercial plasticisers, sublimes at ambient temperatures and so was lost rapidly from cellulose nitrate formulations. The result of plasticiser loss is shrinkage and brittleness. Most PVC formulations contain between 1% and 3% by weight of a lubricant to prevent adhesion of the object to the mould during production. The most common lubricant is stearic acid, which is incompatible with PVC. With time, excess stearic acid migrates to the surfaces of the finished product and forms a white, disfiguring layer.

* *Absorption of liquids or vapours.* Polyethylene is especially vulnerable to absorption of liquids. Tupperware food containers often develop tacky internal surfaces after long-term use due to the absorption of oily materials from contact with foods.
* *Excessive exposure* to high or low temperatures or rapid change in temperature.

Chemical causes include:

* *Oxidation.* Most degradation reaction paths for plastics involve oxygen, either by direct reaction of polymers or additives with oxygen or by the formation of another reactive material from oxygen. Highly unstable ozone (O_3) is the product of the reaction between oxygen and ultraviolet light. It readily adds across carbon double bonds present in synthetic rubbers.
* *Metal ions.* Copper ions act as catalysts to accelerate deterioration reactions of many polymers, particularly synthetic rubbers. Spectacle frames constructed from cellulose nitrate are often more degraded in the area around copper-containing screws and around the wires in the arms, than in the rest of the frame.

In addition to physical and chemical factors, all degradation reactions require energy in the form of heat, light or radiation.

4.1 Degradation of Cellulose Nitrate

The major causes of instability of cellulose nitrate are due to the products of hydrolytic, thermal and photochemical reactions. Degradation of the polymer is autocatalytic, that is, the products of breakdown tend to catalyse a faster and more extensive degradation reaction than the primary processes, if allowed to remain in contact with degraded cellulose nitrate.

One of the first products of thermal deterioration, *i.e.* in the absence of light, is the highly reactive, highly toxic oxidising agent nitrogen dioxide (NO_2), identified by its yellow vapour and distinctive odour. This is formed by cleavage of the N—O bonds joining the cellulose ring that are the weakest bonds in the molecule. Nitrogen dioxide reacts with moisture in air to form nitric acid, known for its ability to corrode metals on contact. Chain scission along the backbone between the cellulose monomeric rings follows, resulting in a considerable reduction in molecular weight.

Cellulose nitrate is particularly susceptible to light of wavelengths between 360–400 nm. Degradation is due to a nitrate ester cleavage in a similar manner to thermal decomposition. At shorter wavelengths, *i.e.* those with higher energy, disintegration of the cellulose ring occurs, causing a rapid decrease in molecular weight. Once started, this process continues even in the dark.

An examination of cellulose nitrate adhesive taken from repairs in cuneiform (dried clay) writing tablets from the British Museum, where the date of the repair was known, suggests that the degradation of cellulose nitrate is retarded substantially by the plasticiser added during manufacture and that such adhesives are stable for at least 30 years. The relationship between loss of plasticiser and reduced stability of cellulose nitrate is also demonstrated by three-dimensional objects. Degradation can be divided into three stages. The first stage involves the evaporation or migration of plasticiser manifested by shrinkage of the object; the remaining material is highly flammable and burns at a temperature up to 15 times higher than that achieved by burning paper. As degradation continues, internal cracks or crazes develop and cellulose nitrate yellows, as shown in Figure 2. In the final stage, crazing known as crizzling is so extensive that cellulose nitrate disintegrates. At this point, its flammability is the same as that of paper.

Some metals, notably copper, accelerate the rate of degradation of cellulose nitrate. Copper screws and arm wires of cellulose nitrate spectacle frames

Figure 2 *Poster made from cellulose nitrate in 1960s showing shrinkage due to loss of camphor plasticiser and cracking*

from the 1930s accelerate degradation of surrounding cellulose nitrate and are, in turn, corroded by the nitric acid formed as a degradation product.

4.2 Degradation of Cellulose Acetate

The major degradation reaction of cellulose triacetate is similar to that of cellulose nitrate, the primary reaction being deacetylation (also known as hydrolysis) during which hydroxyl groups replace acetyl groups ($-COCH_3$) on the cellulose ring. Deacetylation is accelerated by water (usually in the form of moisture in air), acid or base. Because the loss of acetate groups from cellulose acetate results in the formation of acetic acid (CH_3COOH) that gives a distinct vinegary odour to degrading materials, the process is also known as the 'vinegar syndrome'.

Cellulose acetate undergoes autocatalytic breakdown if acetic acid is allowed to remain in contact with the degrading polymer. This is a common situation because the solubility of acetic acid in cellulose acetate is high, comparable with the solubility of acetic acid in water, for example in atmospheric moisture. Degradation does not follow a linear rate, but is slow during the initial, induction period and more rapid after the onset of autocatalysis. As deacetylation progresses, chain scission takes place with bonds breaking between the cellulose monomeric units, dramatically reducing the molecular weight, tensile strength and solubility of the polymer.

Shrinkage, tackiness and increased brittleness due to the migration and subsequent evaporation of plasticiser from between the cellulose acetate chains, is also a frequent cause of degradation. The degradation of some plasticisers has been shown to increase acidity of cellulose acetate-containing materials. Triphenyl phosphate, used as a plasticiser for cellulose acetate since the 1940s, decomposes to form diphenyl phosphate and phenol. Diphenyl phosphate is a strong acid so it is likely to accelerate the deacetylation of cellulose acetate.

4.3 Degradation of Plasticised PVC

In many international museum collections, degradation of plasticised PVC materials, in the form of clothing and footwear, furniture, electrical insulation, medical equipment, housewares, vinyl records and cassette tapes, toys and packaging materials used to store objects, has been detected as early as five years after acquisition.

Degradation of plasticised PVC materials in museums is frequently observed as migration of the plasticiser from the bulk phase to surfaces. From there, plasticiser evaporates at a rate dependent on its vapour pressure. This process may be detected as a tacky feel to the plastic, increasing brittleness and subsequent discolouration of the PVC polymer itself. The mechanism by which

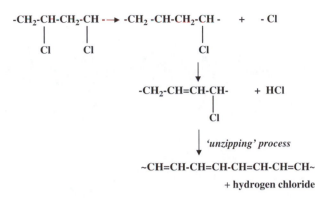

Reaction Scheme 4 *Dehydrochlorination reaction of PVC results in the formation of polyenes and hydrogen chloride*

degradation of PVC takes place is complex. In general, it comprises one major reaction; the evolution of hydrogen chloride (dehydrochlorination process). In addition, cross-linking and chain scission reactions affect the physical properties of the degraded PVC. Cross-linking results in high molecular weight, stiff polymers, while chain scission reduces the molecular weight, thereby increasing solubility.

It is usually assumed that dehydrochlorination starts at imperfections in the PVC structure and starts with the breaking of a C—Cl bond as shown in Reaction Scheme 4. Loss of a chlorine atom is followed almost immediately by abstraction of a hydrogen atom and a shift of electrons in the polymer to form a double bond. The next chlorine becomes allylic, highly reactive and is readily removed. This leads to the progressive 'unzipping' of neighbouring chlorine and hydrogen atoms to form a conjugated polyene system (alternate single and double carbon bonds), accompanied by the formation of hydrogen chloride. As the conjugated polyene system develops, the polymer begins to absorb radiation in the ultraviolet part of the spectrum. After between 7 and 11 repeat polyene units have formed, absorption shifts to longer wavelengths until it is absorbing in the violet, blue and green parts of the spectrum. Each absorption maximum has been found to correspond to a specific polyene length. The rate of degradation can be followed using colour changes from white to yellow to orange to red, to brown and ultimately to black. Dehydrochlorination is an autocatalytic reaction, that is, if the hydrogen chloride produced is not removed from the environment surrounding PVC, dehydrochlorination continues at an accelerated rate.

The rate and extent of deterioration of plasticised PVC and the migration and loss of plasticiser, particularly phthalates, are related. DEHP inhibits the degradation of the PVC polymer, therefore when it either migrates to surfaces or is absorbed by other materials, PVC materials become discoloured, tacky

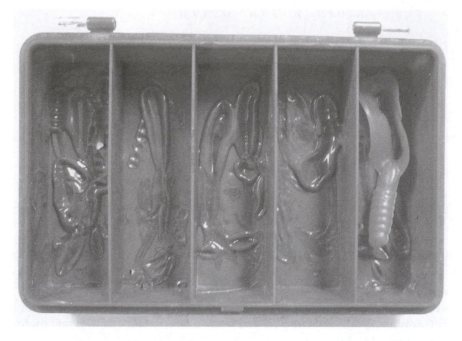

Figure 3 *A polyethylene fishing-box has absorbed plasticiser from plasticised PVC lures and formed polyethylene 'fossils' over 20 years*

to the touch and embrittled. Low-density polyethylene is an effective absorber of oily materials, including plasticisers as demonstrated by a polyethylene fishing-box used to store plasticised PVC lures in Figure 3.

In addition, as esters, phthalate plasticisers are susceptible to hydrolysis when exposed to strongly acidic or alkaline environments. Acid hydrolysis causes the formation of crystalline phthalic acid, volatile 2-ethyl hex-1-ene and 2-ethylhexanol. Acidic environments may develop when the PVC polymer degrades to form hydrogen chloride. Oxygen attack on alkyl groups in the DEHP molecule also results in the formation of phthalic acid.

4.4   Degradation of Polyurethane Foam

Polyurethanes are mainly degraded by oxidation, resulting in discolouration and a loss of mechanical properties. Since oxygen comprises 21% of air, it is difficult to prevent its contact with objects. Polyester-based urethanes are less stable with time than polyether-based urethanes. Polyurethane foams degrade more rapidly than films or fibres since their many cells offer a greater surface area-to-volume ratio over which oxygen can have contact with the polymer. In addition, the processing of foams may involve blowing air through polyurethanes in liquid form, providing conditions favourable to oxidation processes. Exposure to moisture, heat and light leads to polyurethanes' degradation by hydrolysis. Degradation results in chain scission, in which energy breaks polymer chain bonds to create a polymer with two or more shorter chains, manifested by crumbling of foams. Such crumbling often starts at surface skins of foams and, when the surfaces crumble and fall way from the object, fresh, undegraded foam is exposed to light and oxygen, perpetuating degradation to the point of complete failure.

5 CONSERVATION OF PLASTICS

In 1993, a committee of curators and conservators, representing six major museums of modern and contemporary art in the Netherlands concluded that:

- there were no generally accepted methods and criteria for solving the conservation problems of non-traditional objects and modern art;
- there was little insight into the nature and use of modern materials; and
- knowledge concerning the composition and ageing of modern materials was difficult to access.

Deterioration of plastic objects in museums has only been recognised as an important area worthy of research since 1991, while the scope of the problems surrounding the preservation of modern art has only been appreciated internationally as recently as 1998. Little is known about the conservation of plastics compared to other materials found in museums. To date, few conservation practices have been established, and development of new treatments is far slower than most degradation processes. Once deterioration has started, it cannot be stopped or reversed, but only slowed.

Conservators consider two approaches to conservation when planning treatment for any degraded object; *active* also known as interventive, and *preventive* also known as passive. Active conservation treatments are those involving practical treatments applied as necessary to individual objects to limit further deterioration. They include adhering broken sections, cleaning surfaces and filling missing areas to strengthen the objects weakened by deterioration. Although most condition surveys of objects containing plastics conclude that approximately 75% of collections require cleaning, such active conservation practices are still poorly developed. The major cause is the sensitivity of many plastics to organic liquids, aqueous solutions and water itself, particularly if the polymer has deteriorated. Any coating or adhesive that adheres successfully to a plastic surface, must either soften or melt the substrate. This process changes the appearance of the original, which may not be acceptable for works of art. Since this area is so poorly developed and appropriate treatments need to be almost customised to suit individual objects, it will not be discussed further here.

Preventive conservation involves controlling the environments in which objects are placed during storage and display, with the aim of slowing deterioration reactions. Preventive conservation of plastics can either involve the removal or reduction of factors causing degradation, such as light, oxygen, acids and relative humidity, or of the breakdown products that accelerate degradation. If applied successfully, such an approach can help to prolong the useful lifetime of many objects simultaneously, so it can be considered an effective use of resources. Preventive conservation techniques are more likely than active treatments to comply with the ethical practices of professional conservators, primarily that of reversibility, as summarised in the code of practice:

> 'The conservation professional must strive to select methods and materials that, to the best of current knowledge, do not adversely affect cultural property or its future examination, scientific investigation, treatment or function'. *American Institute for Conservation, 1994.*

Since there are no standard guidelines defining appropriate storage environments of museum objects containing plastics as a generic group, and certainly none specifying appropriate treatments for specific types of plastics, those designed to preserve fragile organic materials such as feathers and plant fibres are commonly applied. These include maintaining a stable relative humidity, usually 55 ± 3%, a temperature of 18 ± 2°C, light levels between 50 and 300 lux and good ventilation if the object emits acidic gaseous degradation products. Although successful in reducing the rate of chemical degradation reactions, the use of low temperatures, particularly below 5°C is expensive and inconvenient. The risk of thermal shock to the object and condensation of moisture from air, as a result of its move from cold to ambient temperatures when removed from a fridge (4–6°C) or freezer (−20°C), is also a concern. Dust should be prevented from contaminating the surfaces of objects by covering with acid-free tissue paper or unbleached cotton; however, tacky surfaces should not be covered with absorbent materials.

Specific preventive approaches to conservation have been used successfully to prolong the useful lifetimes of collections containing cellulose nitrate, cellulose acetate, PVC and polyurethanes. The microclimate surrounding plastic objects can be readily and cheaply controlled by introducing an adsorbent or scavenger into their storage areas.

5.1   Conservation of Cellulose Nitrate

Activated charcoal has been used successfully to dramatically reduce the rate of degradation of cellulose nitrate objects. The charcoal is treated with oxygen to create millions of microscopic pores between the carbon atoms, giving a huge surface area density of 300–2000 m^2 per gram. Activated charcoal can be obtained in the form of woven textile, impregnated card or paper, pellets of various size and powder. Although cheaper, pellets and powder are likely to adhere to tacky objects. Its high adsorption capacity for a wide variety of atmospheric pollutants, organic acids, odours and water vapour has been applied to both industrial and domestic situations, including absorbent mats for cooker hoods. Activated charcoal cloth or charcoal-impregnated paper used as packing materials for cellulose nitrate objects will readily adsorb nitrogen oxide degradation products, rendering them unable to contribute to further degradation of the plastic. When all the pores of the active charcoal are filled, no further adsorption is possible and the adsorbent should be replaced. At the time of writing, there has been no research into the volume of gases which can be adsorbed by activated charcoal before exhaustion, and possible desorption that takes place in a museum environment. As a precaution, active charcoal packing materials should be renewed every three years.

The ability of cellulose nitrate to produce corrosive gases that subsequently affect metals in the surrounding area suggests that it should be stored away from sensitive objects in museum collections. Sulfonepthalein indicators, commonly known as Cresol red and purple, detect the presence of acidic degradation products and can be used to reveal the presence of actively-deteriorating cellulose nitrate in a mixed collection. Sulfonepthalein changes colour when in contact with small quantities of nitrogen and sulfur oxides thus allowing early detection of degrading cellulose nitrate. It can be used to draw an outline on a sheet of paper around a degrading object to highlight areas producing acidic emissions. Subsequent prompt removal of the offending artefact helps to preserve the quality of storage for the surrounding collections. The advantage of using sulfonepthalein indicators is their high sensitivity.

5.2 Conservation of Cellulose Acetate

Zeolites, also known as molecular sieves, are a range of hydrated silicates of calcium and aluminium, and have proved effective for inhibiting the rate of deterioration of cellulose acetate. The removal of water from the lattices under intense heating produces pores of particular sizes, enabling the resulting structure to accommodate small gaseous or liquid molecules. Molecular sieve Type 4A seen in Figure 4, has been used to retard the 'vinegar syndrome', which destroys movie film and cellulose acetate objects stored in sealed metal cans or containers, by taking up the acetic acid produced during degradation. Zeolite pellets can be enclosed in semi-permeable polyethylene sachets before placing in the metal cans or containers. With acetic acid effectively

Figure 4 *Zeolite A4 adsorbs acetic acid from degrading cellulose acetate movie film*

trapped by the zeolite and unable to come into contact with cellulose acetate, the risk of autocatalysis is reduced.

5.3 Conservation of Plasticised PVC

The rate and extent of deterioration of plasticised PVC and the migration of DEHP plasticiser are related. DEHP inhibits the degradation of the PVC polymer, therefore when it either migrates to surfaces or is adsorbed by another material, PVC materials discolour, become tacky to the touch and brittle. Degradation may be inhibited by enclosing them in a non-adsorbent material such as glass, containing non-agitated air. Calculations based on the weight lost by model sheets during accelerated thermal ageing, and the rule of thumb concerning the rate of reactions with temperature, indicate that the useful lifetime of plasticised PVC objects may be prolonged more than 10-fold at ambient conditions, by changing the storage environment from a polyethylene bag to a closed glass container. Enclosing plasticised PVC objects, whatever their level of deterioration, is inexpensive to implement, of low practical complexity and allows public accessibility.

This is rather contra-intuitive. Museum conservators and designers are usually advised either to improve ventilation or to include adsorbent materials to remove volatile degradation products from the air space surrounding plastic objects during storage. This would accelerate the loss of plasticiser and thereby reduce the longevity of both new and deteriorated PVC objects.

5.4 Conservation of Polyurethane Foam

Since the degradation of polyurethane foams involves reaction with oxygen, the removal of oxygen limits the extent of the reaction. Ageless® oxygen absorber is one of several similar commercial products originally designed to inhibit the oxidation of foods during transport, and was the first to be evaluated for its suitability for use with museum plastics. It is available as gas-permeable plastic sachets containing finely-divided iron, which oxidises to form iron oxides in the presence of oxygen and water, taking the oxygen from the surrounding environment. The moisture is provided by the presence of potassium chloride in the sachet and is also a by-product of the oxidation reaction. Different grades of Ageless® are available; Ageless® Z is recommended for the preservation of materials with a water content of between 0 and 85%, including plastics. Ageless® Z can be used as a low-cost, convenient alternative to flushing with nitrogen for long-term oxygen-free storage of rubbers and polyurethane foams. It is claimed that Ageless® oxygen absorber reduces the oxygen concentration of an air-tight container down to 0.01% (100 ppm) or less.

Objects are placed in an oxygen-impermeable envelope, such as those prepared from Cryovac® BDF-200 film (a laminate of nylon and polyolefins) or Escal®, a ceramic-coated film into which Ageless® sachets have been introduced. Enclosures are flushed with dry nitrogen to remove any oxygen before being heat sealed as shown in Figure 5. When Ageless® reacts with oxygen, it undergoes an exothermic reaction producing a small amount of heat. In addition, as a by-product of this reaction, a small quantity of water is formed which causes the relative humidity to increase inside the enclosure. The presence of water has no effect on the rate of degradation of rubbers or polyurethanes; however, this is not the case with most semi-synthetic plastics.

Ageless® Eye is the oxygen indicator supplied with Ageless®; it is in the form of a pressed tablet which changes colour from pale pink (less than 0.1% oxygen) to dark blue (greater than 0.5% oxygen). However, Ageless® Eye tends to lose its sensitivity to oxygen after approximately six months, manifested by unreliable colour changes, so an oxygen monitoring device is more reliable.

6 CONCLUSIONS

In conclusion, considerable progress has been made in the field of plastics conservation during the last 15 years. However, it is only with real-time ageing

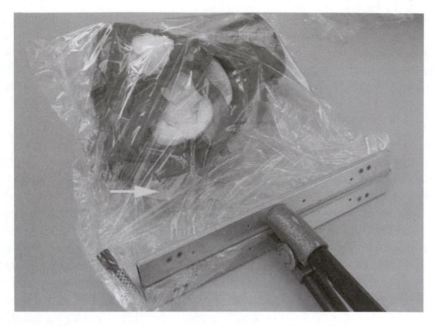

Figure 5 *Rubber gas mask being enclosed with oxygen absorber (indicated by arrow). Bag is flushed with nitrogen just prior to closure by heat sealing*

that the long-term effectiveness of today's treatments may be accurately known. A wider range of adhesives, fillers and cleaning agents needs to be evaluated specifically for application to use with plastics. Without the knowledge of the interaction between such products and historical plastics, progress in developing active conservation treatments will be severely limited.

REFERENCES AND FURTHER READING

J.A. Brydson, *Plastics Materials*, 6th edn, Butterworth-Heinemann, Oxford, 1999.

D.W. Grattan (ed), *Saving the Twentieth Century: The Conservation of Modern Materials*, Canadian Conservation Institute, Ottawa, 1993.

D.W. Grattan and M. Gilberg, Ageless® oxygen absorber: chemical and physical properties. Stud. Conserv., 1994, **37**, 267–274.

European Union Commission, 'Green Paper on Environmental issues of PVC (COM(2000)469FINAL)', 2000, http://www.europa.eu.int/comm/ environment/pvc/index.htm.

U. Hummelen and D. Sillé (eds), *Modern Art: Who Cares?* The Foundation for the Conservation of Modern Art and the Netherlands Institute for Cultural Heritage, Amsterdam, 1999.

International Council of Museums-Committee for Conservation, Modern Materials Working Group, http://www.icom-cc.org.

Macrogalleria is a homepage produced by the Department of Polymer Sciences, University of Mississippi, which clearly describes the chemistry of polymers. http://www.psrc.usm.edu/macrog.

T. van Oosten, Y. Shashoua and F. Waentig (eds), *Plastics in Art, History, Technology, Preservation*, Kölner Beiträge zur Restaurierinung und Konservierung von Kunst- und Kulturgut Band 15, Siegl Munchen ISBN 3-935643-05-5, 2002.

Plastics Historical Society, http://plastiquarian.com have information about the history and production of plastics.

A. Quye and C. Williamson, *Plastics, Collecting and Conserving*, NMS Publishing Ltd, Scotland, 1999.

C. Ward and Y. Shashoua, in *Interventive conservation treatments for plastics and rubber artefacts in the British Museum, 12th Triennial Meeting of ICOM-CC*, vol. 2, pp. 888–893 J. Bridgland (ed), James and James, London, 1999.

CHAPTER 9

Stone

ROBERT INKPEN[1] AND ERIC MAY[2]

[1] Department of Geography, University of Portsmouth, Hampshire, PO1 3HE, UK
[2] School of Biological Sciences, University of Portsmouth, Hampshire, PO1 2DY, UK

1 INTRODUCTION

Degradation of stonework is a major aesthetic and economic problem for historical buildings as well as for contemporary buildings. Oxford University, for example, recently spent £3.2 million renovating Duke Humfrey's Library and other rooms in the Bodleian Library (The Guardian, 22 April 2003, Joel Budd). The nature of degradation can vary from simple discolouration of the surface to the development of micro-scale weathering forms through to potentially structurally-damaging changes. Degradation can be viewed as a general term covering both the weathering of stone, that is, its alteration *in situ*, and the removal of weathering products arising from erosional processes. Often in the scientific literature, weathering and erosion are used interchangeably. Both types of processes are needed to alter stone material and then remove it from the buildings, so conservation practices have looked at means of preventing each set of processes. Evaluating the stage at which degradation becomes a problem requiring remedial measures depends upon its identification as a problem by appropriate building surveyors or architects and the feasibility, often economics, of available conservation measures. Conservators often find an important problem, for conservation must address the variable nature of both the material itself as well as the response of the material to weathering processes and agents. Variation in stone properties and in the nature of the weathering agents that attack it often result in complicated patterns of deterioration across a building, as well as periods of apparently accelerated degradation and periods of seeming quiet. This chapter provides a simple framework for looking at stone degradation and then uses this to assess in greater detail the relative contribution of different factors to the development of distinct types and patterns of degradation. Viewing stone degradation within this framework it becomes easier to identify the types of conservation measures that may be appropriate to particular types of degradation.

2 THE DEGRADATION 'EQUATION'

Degradation can be described by the following equation:

$$D = (f(s, t (MPE)))$$

Where D is Degradation, s and t are space and time respectively, M is material, P is process and E is environment. Any degradation of stonework, both its nature and rate, is the outcome of the variations in space and time of these three inter-related factors, material, process and environment. Degradation at one stage influences the nature and rate of degradation at another stage and so, although general patterns of change may be identified, there is no guarantee as to the precise degradation pathway any particular building or building surface will go through. Each of these factors is looked at separately below, but it is the interaction of the three that produces the complicated nature of degradation as the example of limestone weathering in an urban environment illustrates.

2.1 Material

Stones, or rather rocks, can be divided into three geological types: sediment-ary, igneous and metamorphic. Igneous rocks can be extrusive, such as lavas, or intrusive, such as granite. Differences in the pressures and depths at which these rocks form, the nature of magma as well as the time taken to solidify, influence the type and size of minerals found in igneous rocks. Larger miner-als, in general, result from deeper pressures and longer solidifying times. Igneous rocks contain primary minerals that have differing resistances to weathering processes. The crystalline structure of different minerals, as well as their arrangement, can have a bearing on how prone igneous rocks are to degradation. Silica and silicate minerals, for example, are among the most common rock-forming minerals on Earth, up to 90% of the crust and 75% of all rocks exposed at the surface. Silicon and oxygen atoms can be arranged in a number of ways, some involving the sharing of oxygen atoms with other elements such as magnesium and aluminium, some retaining a tightly packed structure with no sharing of oxygen atoms. The basic arrangements of atoms are as chains, as double chains, as rings, as sheets and as three-dimensional networks. The less oxygen atoms are shared, then generally the more resist-ant the mineral. The different structures are prone to weathering at different

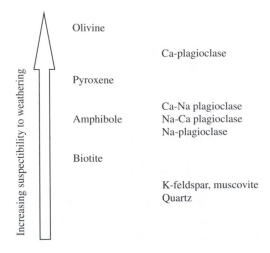

Figure 1 *Goldrich's sequence of mineral stability*

parts of their structures. Pyroxenes, for example, are long-chain minerals and experience chemical weathering along the cleavage planes of the mineral. Goldrich developed a simple ordering of minerals according to their relative susceptibility to weathering, a sequence of mineral stability (Figure 1).

Sedimentary rocks are derived from the weathering products of other rocks. These products are deposited in lakes or the ocean and form layers of sediment. Sedimentary rocks can range from the very finely-bedded limestones to coarsely-bedded and relatively unsorted sandstones. Limestones, such as Portland Stone from the Isle of Portland, are the products of deposition in relatively energy-low environments, such as lagoons. This limestone is composed of microscale ooliths, rounded particles in a finer limestone matrix (micrite). Bedding planes, formed by the different layers of sediment, are present, but they can be relatively inconspicuous compared to the discontinuities produced by changes in the environment within which sediment is deposited, *e.g.* change from lagoonal to shallow sea. This means that certain types of sedimentary rocks may have large discontinuities associated with them, but these can be widely spaced and so be of little concern when the stone is in the building. Indeed in the case of Portland Stone, such large structural divisions mark the boundaries of the different building stone groups (Base, White and Roach beds). Inherent within highly-bedded rocks, such as certain types of sandstones, can be planes of structural weakness. These bedding planes represent more distinct phases of deposition and can be important for the degradation of the stone once it is in the building. Bedding planes can often act as sites for the entry of fluids and as zones of weakness once the rock is stressed.

Metamorphic rocks are either igneous or sedimentary rocks that have been altered by the action of pressure or heat. Marble is an example of a metamorphic rock, being limestone whose crystalline structure has been altered by pressure or heat. Metamorphic rocks tend to be more resistant to weathering agents than the original sedimentary rocks. The almost total loss of pore space and structure reduces the ability of weathering agents to enter the material.

Alteration of any rock, no matter what type, depends upon the ability of agents of weathering to act upon the minerals of which the rock is composed. Alteration tends to be concentrated on surfaces such as mineral grain boundaries. Certain rock properties can either help or hinder this activity. Similarly, once altered, the products of weathering need to be removed to enable further interaction of rock minerals and weathering agents. For any stone the key properties for its weathering behaviour are its mechanical strength, its solubility, its porosity (a surrogate for the ability of agents of degradation to enter it) and the past history of the rock (its memory). Stones which have a low mechanical strength and which are prone to dissolve with high porosity are likely to adjust or change rapidly upon exposure. These stones are likely to be degraded very rapidly as they adjust to their conditions of exposure. Stones that are less porous, less soluble will respond more slowly to their conditions of exposure, thus they will exhibit less degradation and not tend to reflect the prevailing exposure conditions. The latter group may retain more memory, more former surface forms and more weathered products, and so will always be unadjusted to prevailing conditions. This means that a wide range of degradational forms may be present upon a single building. Those present on each surface will reflect the relative degree of adjustment to contemporary conditions and the expression of the underlying characteristics of the stone.

2.2 Process

Processes of weathering have traditionally been divided into chemical, physical and biological weathering. Usually, a range of terms such as haloclasty (salt weathering), gelifaction (ice-induced weathering) and thermal stress are used to distinguish different types of physical or chemical weathering. It may be more appropriate to view these different forms of weathering as being the similar effects of different weathering agents upon the stone. Ice and salt weathering, for example, both operate to induce stresses within the stone and these stresses produce the same effects whatever agent induces them. Variations may exist as to where and how agents operate and so, in consequence, when and where fractures occur. The mechanism by which the fracture occurs, however, is the same – induced stress. It may be more appropriate to rethink chemical and physical weathering as crystal lattice breakdown and stress/strain relationships.

Crystal lattice breakdown refers to the removal of ions from the minerals of which the stone is composed. Removal of ions requires that the chemical bonds between molecules are broken. This requires an energy input, the strength of which will vary depending on the strength of the bonds and the presence of catalysts. Crystal lattice breakdown requires the presence of a substance capable of reaction, as well as a mineral capable of reacting with it. Any chemical reaction involved is an exchange rather than a straightforward removal of ions. There are two basic exchanges, exchange of a proton and exchange of an electron. The former can involve acid solutions, while the latter is the redox (reduction and oxidation) reaction. For removal, it is essential that the reactants can react, that there is sufficient energy to overcome the bonds within the mineral and that the products of reaction are removed from the surface where the reaction occurred. Delays in removal of reaction products means that a layer of unreactive material can build up on a surface, restricting further reactions and so slowing weathering. Where products are removed, such as in flowing water, the surface reaction may be reduced in its effectiveness as the two reactants may not be in contact long enough to permit the reaction to occur.

Brittle fracture refers to the breakdown of stone when stress exceeds the capacity of the stone to deform (strain). The general relationship between stress and strain is illustrated in Figure 2. As stress increases, stone is a material that tends to deform relatively slowly, and by a relatively small amount, and if the stress is removed, the stone suffers no permanent deformation. A point is reached, however, where deformation is permanent, the yield point. Stress and resultant strain beyond this point causes the stone to fracture. There is

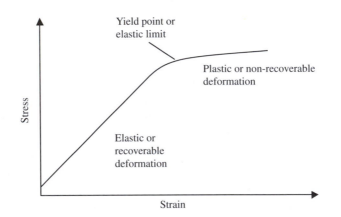

Figure 2 *Stress-strain behaviour of rock. Initially any deformation can be recovered when stress is removed. When yield point is reached further stress produces non-recoverable deformation of the rock, usually fracturing*

irreversible change in the dimensions of the stone. This type of behaviour occurs at every scale. The idea of stress beyond a critical point of deformation can be applied to whole stone facades, or to pores, or mineral grains. For degradation, an important consideration is *what* is causing the stress and *where* in the stone it occurs. Stresses from within the stone require agents to enter and move about the pore structure of the stone. Agents in solution such as salts can move about the stone, although their capacity to induce stress requires them to be activated once within the stone. Similarly, ice-induced stresses can occur from within the stone, but the conditions under which ice expansion can produce damaging strain do not just depend upon it being present. It is the *combination* of material conditions and processes under particular environmental conditions that produces degradation.

2.3 Environment

The environment within which degradation occurs is important for determining both the nature of that degradation and its rate. Environment could be thought of at the macro-scale, such as the climate or atmospheric pollution levels. At this scale general relationships between environment and degradation can be established. Figure 3 illustrates the general model of degradation with atmospheric pollution in Western Europe. Increasing pollution through time accelerates weathering rates and therefore degradation rates rise above a 'natural baseline' level. Within urban areas, there is a spatial pattern to this change as well. At the urban centre, pollution is likely to be higher than in the suburbs and rural areas surrounding it. Zonation of land-use within urban areas can

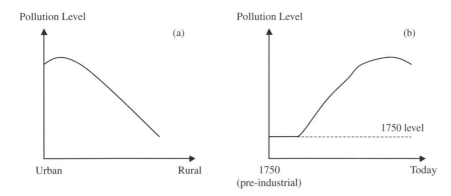

Figure 3 *(a) Spatial pattern of pollution with high values in central urban area, rising in the industrial area, then declining to rural areas. (b) Temporal pattern of pollution increasing from the Industrial Revolution to about 1960s in the UK. This curve will also vary spatially being different for Eastern European and developing nations* (Modified from Cooke, 1989)

alter the pattern from place to place as, for example, industry is regulated to different parts of the urban system. Likewise, as release of pollutants is increasingly from high chimneys, the distribution of pollutants will become more dispersed, but increase in magnitude. This form of release tends to 'export' the effects of pollution from its local area of generation. Models, such as the above at the macro-scale are important in identifying long-term and spatially-wide trends between variables.

Macro-scale patterns do not, however, help in understanding how the effects of environment are mediated at the local level and how they interact with both material and process to produce place-specific degradation forms and rates. At the micro-scale, variations in environment can be vital. The exposure of a surface, for example, can determine the amount of pollution it receives, the magnitude and intensity of rainfall upon it (or if any reaches it at all), as well as the rate and duration of drying. Figure 4 illustrates how an architectural detail could influence the microclimate around it and so produce distinct microenvironments that could accelerate or retard weathering by different weathering agents.

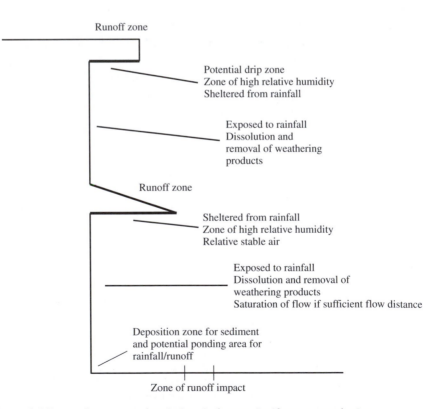

Figure 4 *Microenvironment and variations in factors significant to weathering*

2.4 Limestone Weathering in an Urban Environment: Illustration of the Interaction of the Three Factors

Degradation cannot be pinned solely on one factor – it is the spatially- and temporally-variable interaction of the above factors that produce degradation. An example of how these factors act in combination can be gauged from the example of limestone weathering in an urban or other polluted environment. The material itself, although it may be a natural rock type, is not exposed as a natural rock. Limestone is likely to have been worked before being placed in the environment, at the very least to form it into a shape and size that will fit within a building. This could set up stresses within the stone surface that agents of degradation could exploit. Likewise, the location of a piece of stone in a building, a major determinant of its immediate environment, is not accidental. The location of each stone is designed beforehand, in some cases to take advantage of particular perceived physical properties such as durability. Stones used for copings, for example, tend to be viewed as durable stones as they are expected to have a harder time upon exposure, *i.e.* have greater runoff over their surfaces and greater exposure to wind and driving rain. It cannot, therefore be assumed that all stones are expected to behave in an equivalent manner in a building. In fact, it could be argued that differential degradation is designed into the building. Sometimes, design requirements conflict with the stone properties being exploited. Delicate carving, for example, is generally easiest and most spectacular in stone that is regarded as 'soft' or less durable. The carvings are exposed as much as more durable stone and so tend to degrade at a faster rate. Their propensity for degradation under a given set of conditions is greater.

An initial and important distinction needs to be made between two micro-environments – exposed and sheltered. Exposed surfaces have water flow across their surfaces either from direct rainfall or from runoff or both. However the magnitude and frequency of this flow may vary between surfaces. Sheltered surfaces do not experience any water flow; products of degradation cannot, therefore, be removed from the surface by water flow. This can result in the build-up of a crust of degradation products that could protect the limestone surface from further alteration. On an exposed surface, the build-up of any degradation products is likely to be temporary. Micro-environmental variations in exposure begin to define the type of degradation forms that can be expected to develop on different parts of the building.

For degradation forms to develop, however, it is necessary for reactions to occur that can alter or cause stress in the stone. Delivery of potential reactants to a limestone surface can be by two pathways, wet or dry. Wet deposition occurs when gaseous and particulate pollutants are incorporated into water droplets, falling as rain. This solution is usually acidic; natural rainfall having a about pH of about 5.6 due to weak carbonic acid; urban rainfall falls as low as pH 3.5 from a combination of sulfuric (derived from sulfur dioxide) and nitric acid (derived from nitrogen oxides). The acidic solution will react with the limes (calcium carbonate) according to the reactions below:

Reaction 1.

$$Ca^{2+}CO_3^{2-}(aq) + H_2^+CO_3^-(aq) \rightleftharpoons Ca^{2+} + 2HCO_3^-$$

Calcium carbonate + carbonic acid $\rightleftharpoons$ Calcium + bicarbonate

Reaction 2.

$$Ca^{2+}CO_3^{2-}(aq) + 2H^+SO_4^{2-}(aq) \rightarrow Ca^{2+}SO_4^{2-}(aq) + CO_2(g) + H_2O(l)$$

Calcium carbonate + sulfuric acid $\rightarrow$ Calcium sulfate + carbon dioxide + water

Reaction 3.

$$2(H^+NO_3^-)(aq) + Ca^{2+}CO_3^{2-}(aq) \rightarrow Ca^{2+}(NO_3^-)_2(aq)$$
$$+ H_2O(l) + CO_2(g)$$

Nitric acid + Calcium carbonate $\rightarrow$ Calcium nitrate + water + carbon dioxide

Dry deposition refers to sulfation, the direct gas/solid reaction between sulfur dioxide and limestone. This reaction does require alteration of the sulfur dioxide to sulfur trioxide in the atmosphere as well as the presence of a high humidity level (80% or more) as a catalyst for the reaction. The chemical reaction involved is:

Reaction 4. Basic Reaction

$$SO_3(g) + CaCO_3(s) + H_2O(g) \rightarrow CaSO_4.2H_2O(s) + CO_2(g)$$

Sulfur trioxide + Calcium carbonate + water vapour $\rightarrow$ Calcium sulfate + carbon dioxide

Reaction 5. Reaction Stages

First Stage $\quad CaCO_3(s) + SO_2(g) \rightarrow CaSO_3(s)$

Second Stage $\quad CaSO_3(s) + O_2(g) \rightarrow CaSO_4(s)$

For both these depositional routes, the rate of reaction is dependent upon environmental conditions. Temperature can enhance the reaction rates and for sulfation, water vapour is essential for the reaction to even occur. Similarly, the effectiveness of both reactions depends upon the availability of a reactive surface. For exposed surfaces this is not as easy as it sounds. Surfaces are likely to be at different angles and so runoff velocity will vary. Even for a strong acid, if there is little time to react, then little limestone will be dissolved. A surface cannot be looked at in isolation. One surface will be connected to other surfaces also producing runoff and this will lead to flow between surfaces. The acidity of solutions will change as reaction with the limestone occurs. Likewise, acidity will change as solutions could become diluted by additional runoff volumes. Runoff patterns across a building will therefore influence the rate of dissolution on individual surfaces.

For sheltered surfaces, reactions with pollutants in the atmosphere can only occur if there are no degradation products impeding the flow of pollutants to the limestone surface. If a sulfation crust does form then direct reaction between the atmosphere and limestone will depend upon the rate at which gases can pass through the crust. Some researchers suggest that alteration of limestone to gypsum occurs beneath an existing crust at the limestone/crust boundary. This means that the diffusion coefficients for calcium, sulfate and carbonate ions become important in determining the rates at which crusts can form. Rates of alteration will, however, be affected by the thickness of the existing crust and its permeability, illustrating a negative feedback within the degradation process dependent upon form and so thus indirectly upon the past interaction of material, process and environment.

The development of distinct patterns of degradation in space and time means that there is some predictability about how degradational forms will change. From the same initial starting point, variations in the rate of degradation, in the nature of degradation agents operating on the surface and even in the timing of the actions of these agents, could be used to follow different developmental pathways. The specific pathway followed by a specific surface is almost impossible to predict, particularly when the feedback between one period of development and the next are considered. For a non-calcareous sandstone, some possible developmental pathways are illustrated in Figure 5.

Some modelling of limestone degradation has also been undertaken using the idea of a simple environment-to-weathering relationship. The production of damage functions for limestone relates the ambient concentration of specific atmospheric pollutants to indices of stone damage, such as loss of mass and surface loss. The damage function describes the various relationships between atmospheric pollution and stone damage in Figure 6. The manner in which the stone alters can be linear or non-linear, but the key concept in the figure is that of a threshold above which critical and irreversible damage occurs. Identifying this critical threshold is vital for developing any pollution policy to protect buildings.

It could be argued, however, that the concept of a damage function is problematic for stone. First, all change is irreversible in the sense that stone lost cannot be regained if atmospheric pollution is reduced. The relationship only

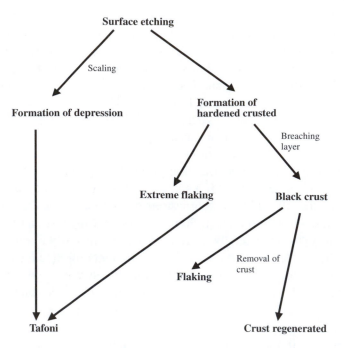

Figure 5 *Potential degradation pathways for sandstone*
(Based on Inkpen and Petley, 2001)

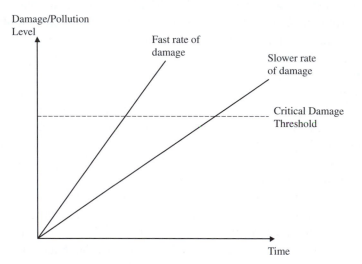

Figure 6 *Damage function for stone. Stone decay proceeds, often at an assumed linear rate, to a critical threshold point. Beyond this point the damage is unacceptable. Unlike some phenomena to which this form of analysis has been applied, reducing pollution does not reverse the effects of damage. Certain suites of processes may operate at a more rapid rate to alter the stone, even under the same pollution conditions, than other suites of processes. This could result in differential damage on a building*

operates in one direction. Without an understanding of how the relationship works, this type of error is possible. Second, deriving the curves and critical thresholds has usually relied on extrapolating laboratory-based studies to field conditions. This often does not take into account the jerky and discontinuous nature of damage. Third, damage functions are often only derived for processes that operate in a relatively simple manner. Figure 6, for example, could be viewed as being a useful expression of the effects of sulfur dioxide on limestone dissolution. Limestone on a building does not, however, only weather by the action of dissolution. Physical weathering by agents such as salts and frost can also cause damage and a different sort of weathering form such as blisters are produced. The development and loss of material from such weathering forms is not necessarily a continuous process nor necessarily related in a simple manner to environmental variables such as pollution levels. This means that it is problematic to represent damage functions as a simple linear relationship between two parameters.

3   MICROORGANISMS AND STONE DECAY

Microorganisms play a crucial role in mineral transformation in the natural environment, notably in the formation of soils from rocks and the cycling of elements such as nitrogen and sulfur. It is therefore not surprising that a wide variety of microorganisms, especially bacteria and fungi, have been isolated from rocks and the stonework of historic monuments and buildings (see Figure 7). The complex interaction of numerous microbial types at a microscopic level in intimate association with the mineral substrate is readily observed, often reaching deeper than 3 cm into the stone. Microorganisms can

Figure 7 *Stone decay and crusts on decorative arches at Portchester Castle associated with high counts of microorganisms*

Table 1 *Microbial activities associated with stone biodeterioration*

Type of activity	Process
Aesthetic	Surface colour change
	Slime production
Biogeophysical	Biofilm formation
	Contraction and expansion of biofilms
	Blockage of pores
	Interaction with salts and water
	Growth/movement through stone
Biogeochemical	Excretion of inorganic acids
	Excretion of organic acids
	Enzyme attack of nutrients
	Chelation of minerals
	Mineral migration

be on the surface or inside stone, as endolithic communities. In some circumstances their long-term surface growth establishes a coloured, varied *patina*, which can sometimes be protective to the underlying stone. Often, however, some types of patina growth leads to damage caused by erosion, biopitting and exfoliation. Research has highlighted a possible role in stone deterioration due to one or more mechanisms: their presence as undesirable *surface growths* (aesthetic), *mechanical damage* (biogeophysical change) by biofilms or penetrating hyphae and *corrosive effects* (biogeochemical change) due to metabolic activity (Table 1). Scientific investigation of these effects can present severe problems with objects of cultural value. Phototrophic organisms such as higher plants, lichens and mosses, together with algae and cyanobacteria, cause obvious surface effects. The impact of most bacteria and fungi is more difficult to appreciate and separate from purely physical and chemical phenomena that are acknowledged threats to the integrity of building stone.

Biodeterioration of stone is rarely associated with one group of microorganisms; weathering stone may support a balanced community whose members co-evolve with time to enable recycling of essential elements for activity and growth. Damage may thus be gradual through slow growth (biogenic *drift*) or be sudden and harmful stimulated by a dramatic change in environment, moisture or nutrients (biogenic *shift*).

Microbial colonisation of building stones is characterised by a biological succession. Colonisation and conditioning of fresh stone by predominantly phototrophic types, which use light for their energy (*cyanobacteria, algae, lichens*), will enrich the stone so that *fungi, bacteria and actinomycetes* can grow on accumulated organic matter, derived from dead cells and trapped debris. Some types obtain their energy from inorganic chemicals (*sulfur-oxidising and nitrifying bacteria*) and may become significant wherever inorganic nitrogen or sulfur compounds are available.

3.1 Light-dependent Microbial Growths

Algae are photosynthetic and develop on stone when light is present. Algae often foul or stain stone surfaces without surface changes. However, algal communities on stone are often embedded in surface slimes together with heterotrophic bacteria. These undergo considerable volume changes through repeated wetting and drying and this has the effect of loosening the stone particles to promote decay. The main contributions to decay are to encourage water retention and facilitate succession by more aggressive microbes although corrosive acids have also been shown to be produced on marble and limestone. *Cyanobacteria* are bacteria that can colonise stone and produce aesthetic changes due to stains, coloured biofilms and incrustations. Since they are light-dependent, they are considered to be pioneers in the colonisation process. Their tolerance to desiccation, water stress and varying light intensities help to explain their frequent occurrence on stone surfaces.

Lichens are 'microbial' in the sense that they have algal and fungal cells in close association, forming a visible *thallus*. They can tolerate extreme dehydration and nutrient limitation in the absence of algae or mosses although they are sensitive to air pollution. Growing slowly on (epilithic) and in (endolithic) stone, they are undoubtedly the cause of damage through mechanical and/or chemical means. Deterioration can be caused by the mechanical effect of substratum-penetrating fungal hyphae (bleaching, blistering or sloughing), excretion of oxalic acid and complexing and leaching of stone minerals by chelation.

3.2 Organics-dependent Microbial Growths

Fungi are associated with the deterioration of stone and the mechanism of attack is thought to be both mechanical, due to growth of filamentous hyphae, and chemical, as a result of acid secretion. Fungal filaments penetrate many millimetres into porous stone. One group of fungi isolated from stone are the rock-inhabiting fungi consisting of black yeasts and meristematic fungi, a heterogeneous group of black-pigmented fungi that survive extreme conditions of humidity and sunlight. The latter group are more ubiquitous and widely distributed in soil and organic material.

Actinomycetes are filamentous bacteria that are often observed on stone surfaces and a large range of types has been isolated from stone. Like fungi, the filaments cause mechanical damage to stone by penetration and development of an extended web of hyphae in the stone (mycelium). These bacteria can produce patches of growth on stone particles and around stone pores, often interacting with salt crystals, which can further enhance the deteriogenic effects of salts.

Heterotrophic bacteria degrade organic matter and can be isolated in large numbers from decaying stone. All stonework probably possesses sufficient organic matter from soil, dust and dirt to sustain heterotrophic activity. Many stone bacteria have a preference for low concentrations of organic nutrients and may even be oligotrophic in nature. Population activity has been related to seasonal and climatic changes, and isolated bacteria can produce acids that cause morphological alteration of the stone surface and elution of minerals.

3.3   Inorganics-dependent Microbial Growths

Sulfur-oxidising bacteria convert inorganic sulfur compounds to sulfuric acid that can cause severe damage to mineral material. *Thiobacillus* species have been implicated with concrete corrosion in the Melbourne and Hamburg sewer systems due to sulfuric acid formation. However, a role in stone decay is less certain since sulfuric acid and calcium sulfate in stone can originate from the direct action of atmospheric pollution and acid rain.

Nitrifying bacteria oxidise inorganic nitrogen compounds for energy and generate acidic end-products, namely, nitrous acid or nitric acid. Nitrifying bacteria can be isolated from stone material but a role in stone decay will be favoured in those buildings where there is an obvious source of ammonia or nitrite. Ammonia may be carried onto stone in dust as ammonium salts while nitrite can originate from the automobiles, soil or industry. Nitrifiers often exist in a biofilm on the surface and within the pores of the stone and *Nitrosomonas, Nitrospira* with *Nitrosovibrio* are commonly isolated.

3.4   Stone Colonisation and Biofilms

The stone ecosystem is subject to harsh environmental change, especially temperature and moisture, exerting extreme selective pressure on any developing microbial community. The complex consortium of micoorganisms that exists on weathered building stone at any given time is the result of ecological successions and interactions that directly relate to fluctuating substrate availability and environmental conditions. Initially, the mineralogy and structure of stone in relation to its capacity to collect water, organics and particles will control its predisposition to biodeterioration, or bioreceptivity.

The ability of the stone-colonizing microflora to cover and even penetrate material surface layers by the excretion of organic extracellular polymeric substances (EPS) leads to the formation of complex slimes, or biofilms, in which the microbial cells are embedded. Phototrophic organisms usually initiate colonisation by establishing a visible, nutrient-rich biofilm on new stone from which they can penetrate the material below to seek protection from high light intensities or desiccation. Stone EPS trap aerosols, dust and nutrients, minerals,

Figure 8 *Biofilm on weathered stone*

and organic compound complexes; and take up water from air and release it under low RH conditions. Stone moisture and nutrients are thereby increased while porosity, water-uptake capacity and evaporation are reduced. Notably rich and homogeneous biofilms, composed mostly of bacterial rods, are often observed on weathered stone substrates from sheltered areas (Figure 8).

Microorganisms may degrade stone mechanically, chemically and aesthetically through metabolic activities and biomineralisation processes in these biofilms. The mechanical stress induced by shrinking and swelling of the colloidal biogenic slimes inside stone pores may damage stone and it may cause changes in the circulation of moisture to further enhance chemical dissolution and mineral loss from stone.

3.5 Interactions of Microbes with Stone Salts

Salts acting on their own are very important decay agents and can attack stones, mainly mechanically in pore spaces during RH and temperature changes. Efflorescences present a niche for salt-tolerant and salt-loving bacterial populations which are osmotically well-adapted to an extreme existence (typical of an ancient group of bacteria known as the Archaea). High concentrations of sodium chloride and magnesium sulfate (up to 25%) may be appropriate for studying bacteria from efflorescences on stone monuments. It has also been shown that microorganisms can enhance the physical or chemical

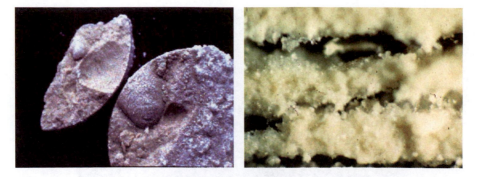

Figure 9 *Stone discs showing exfoliation after treatment with salts and mixed microbial populations*

processes by interacting with salts in stone. When limestone has been subjected to both microbial and salt weathering, under different temperature/wet/dry cycling regimes, weight loss was higher with microbes alone (7.7%) than Na_2SO_4 alone (4.9%) but the two agents together more than doubled the additive effect and caused extensive exfoliation and fissure formation (Figure 9). Thus, by interacting with the effects of the salt, microbial biofilm growth can enhance water content and enhance physical, mechanical pressures on stone during wet/dry cycling.

4 CONSERVATION PRACTICES

A number of definitions of conservation have been developed in relation to stone degradation, but most have their origin in the *Venice Charter* of 1964 that tried to establish guiding principles for the preservation and restoration of ancient buildings. The Charter highlights the importance of contextualizing the building to be conserved and emphasises the need for recourse to scientific methods for conservation and preservation practices. It also provides guidelines for practices to be employed, stating these should not be out of scale with the setting of the monument and that restoration should be:

> *to preserve and reveal the aesthetic and historic value of the monument and is based on a respect for original material and authentic documents*

By 1994, the Nara Conference on Authenticity built upon the *Venice Charter* to define conservation as:

> *All operations designed to understand a property, know its history and meaning, ensure its material safeguard and, if required, its restoration and enhancement.*

In these documents, and in practice, there can be a tension between the terms, or rather interpretation of, preservation and conservation. The Venice Charter, for example, states that removal of material should be viewed as the last option in any restoration work. This can be extended to a general interpretation of preservation as a philosophy that puts a premium on retaining any monument as it stands. It is a philosophy that is static and rigid in its view of a monument – it should be frozen as it is now, as a snapshot in time. Conservation could be viewed as a more fluid and dynamic philosophy that sees a monument as a dynamic and holistic entity, something that changes, that is part of its context and informed by that context. Conservation accepts a balance needs to be struck between preservation and change. A building is designed, built and maintained using the methods available at the time. Building practices change over time and repair work is probably carried out using best, or appropriate, contemporary practices. This means that the building as it is now is unlikely to be a pristine example of traditional practices of a given period. A building is more likely to be a reflection of building practices of different periods.

An important question arises: *what is it that is being conserved?* A building can be thought of in a number of ways. It is constructed of a variety of materials in varying states of degradation. A building is more than individual blocks of stone, it is also a representation of an architectural plan. Which of the above is the focus of conservation and can the two be separated? The distinction is vital, because if the material used is not integral to the representation of the design, then why preserve the material? There may be no easy answer to this question as the design and the material are interwoven, but understanding why particular materials were selected, and how they might be improved or replaced without affecting the representation of the building, is important within conservation.

Conservation does not imply a single approach or method. There are a range of techniques that can be employed. Each technique should, however, be assessed on the basis of a set of criteria before any conservation practices are used. These are reversibility, compatibility, re-treatability and effectiveness. Reversibility refers to the ability to reverse the effects of any treatment. Often the changes that treatments are designed to produce are intended to be irreversible, such as the removal of black crusts by washing. Compatibility refers to the neutral or positive impact of treated materials in a building with the untreated materials. It implies that treated materials should have physical and chemical properties compatible to those of the untreated material. This should ensure that physical and chemical interactions between the two are not harmful to the 'traditional' materials. This criterion also means that it may be inappropriate to use the most modern methods if the results could have a harmful effect on the rest of the building. This could be viewed as implying that there are different tolerance limits for altering material properties by conservation treatments that depend on the context of the other materials. Re-treatability refers to the possibility that new or different treatments can be applied without any harmful effects on the building. Assessing re-treatment implies knowledge of the effects of treatments on stonework and their potential interaction with different and new types of conservation treatments. It is difficult to assess the interaction of existing treatments with each other unless sufficient time has elapsed and good monitoring records have been kept. Assessing the potential impact of new treatments is even more problematic as there is no possibility of obtaining data of interactions over time. Treatment effectiveness can be defined in terms of slowing the rate of degradation and increasing material durability. Both these concepts are difficult to judge without sufficiently long exposures and monitoring. All the above emphasise a very conservative view of conservation; most require long exposures and detailed records to enable any judgement, other than anecdotal, to be made.

4.1 Identification

The first step in any conservation project should be identification. It is essential to try to identify the nature of degradation and its severity. This will influence the type of conservation practices undertaken. Of particular significance is establishing the spatial extent of the types of degradation. This identifies which treatments are needed and where they are to be applied on the building. A key part of identification is developing an appropriate classification scheme. There are a number of different schemes in existence, many of which rely upon visual characteristics of the degradation form for assigning features to specific classes. A number of these schemes also assume that there is a specific and known link between particular decay forms and a set of, or even single, decay agents or processes. This link is usually assumed, rather than established by theoretical understanding or even based on a vast amount of empirical data. Table 2 presents a simplified scheme for identifying and distinguishing different forms of degradation based on a limited set of visual characteristics. Although not ideal, the scheme does not assume a simple process-form relationship and so does not assume that certain forms are produced by a particular process or a set of processes.

4.2 Prevention

One of the best means of slowing degradation is to prevent the agents that cause it from reaching the building in the first place. Prevention is, however, also one of the most difficult objectives to achieve. Preventing atmospheric pollutants from reaching a building in an urban area would be possible if atmospheric pollution levels were reduced. Reducing levels, however, requires

Table 2 *Simple classification of weathering forms on buildings*

Type	Description
Detachment and disintegration	Alteration of surface leading to detachment and removal of material, but no major raising or lowering of surface, *e.g.* crumbling, flaking and spalling
Deposition and staining	Change in colour of surface
Relief forms	Alteration of surface resulting in major raising or lowering of surface. Major is defined here as noticeable change in surface height relative to the original surface as can best be determined by the investigator, *e.g.* blistering, pitting and back weathering or hollowing out of forms
Cracking	Single or multiple cracks visible
Weathering products	Products of weathering found on surface or recesses. These are products of weathering and do not refer to the texture of the surface, *e.g.* grains or powdered material
Repair and replacement	Material visibly distinct from main body of surface that observer can identify as repair or replacement work, *e.g.* plastic repair

management of all the activities that produce pollutants, a virtually impossible task. Even limited management of these activities is difficult to justify given the relatively poor understanding of the level at which pollutants can damage stone (see the dose functions mentioned above). Current policies of pollution reduction could have the side-effect of reducing pollutant delivery to stone surfaces. The impact of these reductions upon stone degradation may not, however, be immediate. The so-called 'memory-effect', the weathering of stone by the salts deposited by past pollution, will cause degradation above that expected from current pollution levels alone. This additional degradational effect will continue to operate years after pollution levels have been reduced.

4.3 Barriers to Agents of Degradation

Another preventative method involves the use of barriers on or within the stone that prevent weathering agents from reaching and interacting with the stone. Limewash, a traditional method of protecting building stone is a barrier treatment. Coats of limewash are applied to a building and act as sacrificial layers protecting the underlying stone. The limewash weathers preferentially relative to the underlying stone, as it is the surface upon which reactants land. Removal of the limewash, however, means that the sacrificial layer needs to be renewed frequently, requiring a monitoring and maintenance programme for the building. Treatments such as infilling damaged areas with silica-based polymers can also prevent agents of degradation reaching the surface. These treatments do not act as a sacrificial layer, but rather block the path of any agent to the reactive stone surface. The surface of the stone may be protected from external agents, but unless the protective layer is able to let internal agents through it, there is a possibility that the stone will be damaged below the protective surface. At the boundary between the stone and protective layer, agents such as salts may, if present, build up and eventually create such high stresses as to cause a catastrophic failure of the stone surface.

4.4   Removal of Degradation

Where degradation has occurred the products of degradation can be removed, but, importantly, the question of whether it is essential to remove these products needs to be asked. Where a surface is discoloured through, for example, the accumulation of dust or by the development of a sulfation crust, removal of the dust may require contact with the surface by abrasive agents, such as brushes or by water. The process of removal may have an impact upon the stone surface causing alteration. Inputs of large quantities of water, for example, could result in water penetration into the stone and reactivation of salts stored relatively harmlessly in the stone (the 'memory-effect'). Likewise, abrading a stone surface could remove both the accumulation layer as well as 'sound' stone underneath. Unless a highly-selective removal method is employed it is likely that the stone surface remaining after treatment will have suffered some physical alteration at the microscale.

A common method of removal is the use of poultices. The term has been adapted from its original use in medicine where it indicated the application of a treatment that removed infection from a body. For building materials, the application of the treatment enables contaminants or staining to be 'drawn' out of the material. A poultice has to have the ability to mobilise the contaminants and then actively encourage the movement of the contaminants into the poultice. Common poultice media are clay and paper fibres mixed with water. The clay enables the mixture to adhere to the material, while the water acts as the medium for mobilisation. Salts are drawn into the poultice by capillary action as the poultice dries, although it may take a number of repeat treatments to removal all the contaminants. Specific additives may be mixed into the poultice, such as EDTA, which will target particular contaminants that are not water-soluble, in the case of EDTA, copper and iron stains. Using poultices with additives usually requires a moist surface for application to prevent deep penetration of the additives as well as through washing of the surface after application. Although poultices can be used for removing contaminants from whole surfaces, other treatments such as water spray are often cheaper. Poultices do, however, provide a degree of control in application of active agents that makes them very useful cleaning techniques for delicate or intricate sculpture and stonework.

Bioremediation for buildings. The term bioremediation covers a range of processes that utilise microorganisms to return contaminated environments to their original condition. As discussed above, deterioration of building stone begins from the moment it is quarried as a result of natural weathering processes. Other factors, sometimes acting synergistically, including crystallisation of soluble salts, pollution and biological colonisation can accelerate natural deterioration. Internal pressures created by crystallisation, hydration and thermal expansion of salts are a significant cause of damage to stone. Accumulation of sulfates, derived from the oxidation of sulfur dioxide, and nitrates, from oxides of nitrogen present in the atmosphere (N_2O, NO, N_2O_3, NO_2, N_2O_5), is of particular concern. Whatever the cause of stone deterioration, buildings require remedial measures to stabilise the surface layer and prevent further loss.

Conventional techniques for the cleaning and conservation of stone in buildings have a number of disadvantages: they can cause colour changes in the stone; excessively remove the original material; or adversely affect the movement of salts within the structure of the stone. While microorganisms have usually been associated with detrimental effects on stone, affecting mineral integrity or exacerbating powerful physical processes of deterioration, there had been growing evidence that some types can be used to reverse the deterioration processes on historic buildings and objects of art. Bacteria, such as *Pseudomonas and Desulfovibrio*, have shown potential to remove harmful salts such as nitrate and sulfate by denitrification and sulfate reduction and to mineralise organic residues or pollutants like carbohydrates, waxes or hydrocarbons which commonly occur in crusts on stonework. Research suggests that bioremediation, may offer a safe, viable alternative to conservators and restorers working to reverse or prevent further damage to stone buildings.

On the surface of the stone, particulate matter from the atmosphere can combine with gypsum to leave unsightly black crusts containing a complex mixture of aliphatic, aliphatic and aromatic carboxylic acids and polycyclic aromatic hydrocarbons. The removal of black crusts is problematic for conservators since conventional cleaning techniques may remove a portion of the underlying stone. Recently, microorganisms have been used to remove sulfate from black gypsum crusts. Sulfate-reducing bacteria (SRB) are able to dissociate gypsum into Ca^{2+} and SO_4^{2+} ions, and, in the absence of oxygen, the SO_4^{2+} ions are then reduced by the bacteria to a gas, H_2S. In some cases, the Ca^{2+} ions have been shown to react with carbon dioxide to form new calcite. *Desulfovibrio desulfuricans* and *D. vulgaris* are two types of bacteria that have been applied to gypsum crusts and used to successfully remove black crusts from marble.

4.5   Control of Microbial Growths using Biocides

The term 'biocide' refers collectively to fungicides, algicides and bacterio-cides. They may serve to either inhibit growth/metabolic activity while present, with effective metabolism being restored upon removal (*biostatic*), or cause a lethal effect, through irreparable damage to target cells leading to cell death (*biocidal*). The latter effect may be due to action against cell proteins, consequently biocides are potentially harmful to humans and their toxicological properties should be understood so that appropriate risk assessments can be carried out. Unfortunately most biocides have not been designed specifically for object conservation and therefore adverse reactions and compatibility have rarely been considered.

The ideal features of a biocide are that it will have a wide spectrum of activity, kill target organisms but pose no harm to humans, have a long effective life and cause no changes to stone. Unfortunately no candidate biocide fulfils these criteria. Consequently there are many factors to be considered when choosing a biocide for use in stone conservation and restoration. Clearly the important and obvious factors are the range of effectiveness against target organisms and any possible unwanted toxicity to humans. The potency of biocides may be influenced by the ability of microorganisms to oxidise or modify them, possibly in conjunction with UV light. It must also be considered that, although vegetative growth of microorganisms may be sensitive to biocides, very often their spores are resistant. Thus the elimination of obvious surface growths may be followed very quickly by recolonisation, as spores germinate and grow on dead organic matter residues when the active biocide concentration declines.

Biocides in use.   There are 5 main types of biocides that are active against microbial growths:

1. Quaternary ammonium compounds
2. Amines
3. Chlorophenols
4. Phenoxides
5. Metals.

Biocidal action is best developed in the quaternary ammonium compounds, which are surface-active and can be bacteriocidal or bacteriostatic depending on concentration. They have a wide spectrum of activity against microorganisms, although they are less effective against one important class of bacteria, the Gram-negative bacteria. Biocide treatments containing copper have been reported as having the greatest chance of success in controlling lichen growth. Nevertheless, whenever using biocides for controlling microbial growths, positive identification of the target organisms is critical. This allows the selection of an appropriate chemical for the problem organism. Since few biocides have a wide spectrum of activity or resistance may arise, combinations or rotations may be necessary to achieve the desired effect. Toxic chemical washes have certainly been used to eradicate or remove unsightly biological growths from stone but these growths may be succeeded by mosses and higher plants with greater damage potential.

Factors to consider. Success with biocides depends very often on the residual effective concentration of biocides after application. This, in turn, depends strongly on the nature of the stone and its growths. The density of any existing, or previous, growths or dirt will directly affect penetration to target populations and may cause inactivation, via adsorption reactions with the biofilm, or even biodegradation. The stone itself may cause losses or deactivation by adsorption to stone components; its porosity may cause it to act as a reservoir (if low) or barrier (if high). In addition, leaching by rainwater, evaporation, photo- or chemical oxidation will all reduce the concentration of the biocide that remains to do its job. Maximum impact requires binding with minimum deactivation. The persistence of biocides will be affected by environmental factors such as pH and temperature and the type of environment they are used in, particularly its microclimate and exposure. Re-application is usually necessary.

The desired biological action of biocides against unwanted growths on stone, of course, also means that these chemicals pose a threat to the natural environment if they cannot be restricted to the stone substratum. Thus the chemical and physical behaviour, especially their solubility in water and capacity to react with stone components, has an important effect on whether there is a risk of environmental pollution. The interactions with the stone, or indeed previous chemical treatments, could also lead to colour changes or even mechanical damage through crystallisation.

4.6 Consolidation

Removal of weathering products may enhance the aesthetics of a building, but removal may be detrimental to cohesion of the structure of that part of the building. Statues, for example, may have sulfate crusts removed, but the crusts may be the only solid component of that part of the statue. Calcium carbonate may have long ago been replaced by calcium sulfate and so the details of a statue's face could be composed solely of weathering products.

Biocalcification. An alternative to chemical consolidation of stone is the exploitation of a common phenomenon found in living organisms known as biomineralisation, which produces shells in animals. Microorganisms can also be isolated from the environment or stone surfaces that are able to precipitate

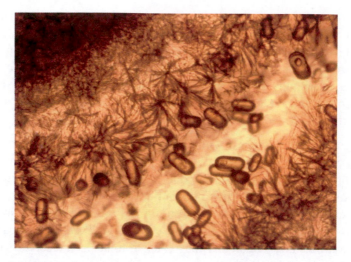

Figure 10 *Calcinogenic bacteria in laboratory culture, showing calcite crystals developing within colonies*

$CaCO_3$ in the form of calcite. Although biomineralisation has been observed for many years, the potential for its use in stone consolidation has only been explored relatively recently. Biocalcifying bacteria have been used to successfully promote carbonatogenesis on the surface of limestone buildings, statuary and monuments. This newly-formed carbonate is often resistant to mechanical stress, possibly due to the incorporation into crystals of organic molecules, produced as a result of bacterial metabolism.

Bacillus cereus has been shown to protect exposed mineral surfaces by the formation of sacrificial layers of calcite, vaterite or aragonite crystals, which may be dissolved in a polluted environments but can be renewed when necessary. Other non-sporing bacteria such as *Micrococcus xanthus* have also been shown to produce calcite ot vaterite crystals which strongly adhere to the original stone and production can be controlled by changing the environmental conditions. Such bacteria precipitate calcium carbonate in their immediate environment (Figure 10) and encrust cells in the process of carbonatogenesis (Figure 11).

It is now understood that application of consolidants to stone surfaces may hinder the movement of salts that then accumulate, leading first to unsightly discolouration but ultimately physical damage. While some reservations exist regarding the application of bacteria to stone, calcium carbonate generated by bacteria could offer a solution to this problem because the production of a layer of calcium carbonate on calcareous stone that does not block the natural pore structure, allowing free passage of soluble salts through the stone.

Figure 11 *Encrustation of bacteria cells applied to stone during biocalcification activity*

Bioremediation into conservation practice. Bioremediation is less harsh than the use of environmentally-toxic chemicals or aggressive mechanical procedures, which should be considered to be destructive methods. Research into the practical application of bioremedial technologies has recently progressed through the work of two European Community-funded projects. BIOREINFORCE successfully demonstrated that dead cells from active biocalcifying strains showed a much higher and/or faster production of $CaCO_3$ crystals than less active strains. By identifying the genes responsible for crystal formation in bacteria, the project aimed to biologically produce chemicals that induce calcification on stones without using living bacterial cells. BIO-BRUSH (www.biobrush.org) linked the mineralisation processes that remove stone crusts to consolidation by biomineralisation. Bacteria were applied directly to stone surfaces using techniques with a low degree of risk to both environment and conservators. Multiple short-term applications of aerotolerant sulfate-reducing bacteria within an appropriate delivery system was very successful in removing black crusts from marble both in the laboratory and *in situ* on buildings. A new mortar system incorporating nitrate-reducing bacteria capable of removing nitrates was also developed. Biocalcifying bacteria were shown to deposit a calcite layer without significant reduction in porosity or subsequent growth of contaminating microorganisms.

4.7 Replacement

When consolidation methods fail or are inappropriate due to costs or conservation policy, then, usually as a last resort, replacement of material is required. Replacement is problematic in some buildings because it is not possible to replace like with like. Original quarry sources may have long since ceased operation. It is often difficult to identify and acquire stone types of similar physical and chemical properties. Even where such matches are available, very little may be known about the weathering behaviour and, importantly, alterations in colour of material upon weathering. For a period of time the replacement stone may be visually sharply distinct relative to the rest of the fabric of the building. Portland Stone, for example, when initially emplaced is a bright white colour. As weathering begins, the colour alters and it is hoped that it will eventually blend in with the existing fabric of the building. Unfortunately, the weathering behaviour and associated discolouration of stone is not clearly understood. Additionally, the conditions under which stone weathered in the past are not the same as the conditions under which it weathers today. Within the UK, for example, pollution control has severely reduced smoke and sulfur dioxide levels in urban areas. This means that limestone exposed today will not experience high smoke and sulfur dioxide levels in its initial stages of weathering. What significance, if any, this different weathering history will have for the appearance of the stone is unknown. Choice of appropriate replacement stone is often based on the experience of the architect or stonemason in charge of each project. The stone that is used for reference is likely to have been weathered under environmental conditions different from those in which the replacement stone is exposed. Importantly, emplaced stone is likely to have been weathered in polluted urban environments of the late 19th and mid-20th centuries.

5 CONCLUSION

Stone decay is a complicated process involving the interaction of a number of different factors. Alteration of any rock, no matter the type, depends upon the ability of agents of weathering to act upon the minerals of which the rock is composed. A wide range of degradational forms may be present upon a single building. Those present on each surface will reflect the relative degree of adjustment to contemporary conditions and the expression of the underlying characteristics of the stone. Processes of weathering have traditionally been divided into chemical, physical and biological weathering. The environment within which degradation occurs is important for determining both the nature of that degradation and its rate.

The interaction between microorganisms and stone minerals demonstrates metabolic diversity and versatility, combined with remarkable tolerance to extreme environmental conditions. Biocides can be used to eliminate some microbes but they should have extended activity and a wide spectrum of action. They should be checked for detrimental effects on the stone, either through colour change or internal crystallisation. The use of microorganisms to restore building stone by bioremediation is a new technology that offers a different approach for conservators to supplement, not replace, existing conservation technologies. It is not yet readily available to all, but research suggests a viable alternative technique for conservation of stone.

Degradation of stone cannot be pinned solely on one factor – it is the spatially and temporally variable interaction of these factors that produce degradation. It is the combination of material conditions and processes under particular environmental conditions that produces degradation.

6 SUGGESTIONS FOR FURTHER READING

As a general introduction to the problems of stone degradation, its causes and patterns across the United Kingdom, *Crumbling Heritage* by Ron Cooke and Alan Webb (1995) is a useful start. The book is written with the non-specialist in mind, but contains a great deal of information about the scientific analysis of processes of stone degradation and techniques used to determine patterns of degradation across the UK. Despite its age, the classic text on stone degradation by Schaffer (1932) is still a good initial starting point for understanding the processes of degradation and potential treatments. A useful text on techniques of stone conservation is provided by Ashurst (1988). This book is a guide to practical conservation techniques and is part of a series of such books published through English Heritage. Lastly, Baer and Snethlage (1997) text *Saving Our Cultural Heritage* is a series of papers taking a more detailed look at the philosophy and practice of stone conservation from a more European perspective. An interesting website for stonemasonry is that of the Masonry Conservation Research Group at the Robert Gordon University in Aberdeen (http://www2.rgu.ac.uk/schools/mcrg/mcrghome.htm). This site contains a great deal of information on processes of stone degradation as well as treatment of stone. In addition, it has an extensive number of links to other conservation groups and sites of interest over the Web.

REFERENCES AND FURTHER READING

J. Ashurst and N. Ashurst, *Practical building conservation: English Heritage technical handbook*, vol 1, Stone masonry, Gower, Aldershot, 1988.

N.S. Baer and R. Snethlage (eds), *Saving our cultural heritage: The conservation of historic stone structures*, Dahlem Workshop Report. Wiley, Chichester, 1997.

S. Cameron, D. Urquhart, R. Wakefield and M. Young, Biological growths on sandstone buildings: control and treatment. Technical Advice Note 10, Historic Scotland, Edinburgh, 1997.

R.U. Cooke, *Crumbling heritage? Studies of stone weathering in polluted atmospheres*, National Power, Swindon, 1995.

O. Guillette, Bioreceptivity: a new concept for building ecology studies, *The Science of the Total Environment* 1995, **167**, 215–220.

R.J. Inkpen and D. Petley, Fitness spaces and their potential for visualising change in the physical landscape, *Area*, 2001, **33**, 242–251.

Masonry Conservation Research Group, http://www2.rgu.ac.uk/schools/mcrg/mcrghome.htm at the Robert Gordon University. Accessed 9 June 2003.

E. May, S. Papida and H. Abdulla, Consequences of the microbe-biofilm-salt interaction for stone in monuments. In: R. Koestler (ed.), *Art, Biology and Conservation: Biodeterioration of Works of Art*, 2003, In Press, Metropolitan Museum of Art, New York, 2003.

E. May, F.J. Lewis, S. Pereira, S. Tayler and M.R.D. Seaward, D.A. Allsopp, Microbial deterioration of building stone – a review. *Biodeterioration Abstracts*, 1993, **7**(2), 109–123.

S. Papida, W. Murphy and E. May, Enhancement of physical weathering of building stones by microbial populations. *International Biodeterioration & Biodegradation*, 2000, **46**(4), 305–317.

G. Ranalli, M. Chiavarini, V. Guidetti, F. Marsala, M. Matteini, E. Zanardini and C. Sorlini, The use of microorganisms for the removal of sulphates on artistic stoneworks. *International Biodeterioration & Biodegradation*, 1997, **40**, 255–261.

R.J. Schaffer, *The weathering of natural building stone,* Department of Scientific and Industrial Research, Special Report 18. HMSO, London, 1932.

S.T. Trudgill, Weathering overview – measurement and modelling. *Zeitschrift fur Geomorphologie N.F.* 2000, Suppl. Bd. 120, 187–193.

A.M. Webster and E. May, Bioremediation for buildings. *Trends in Biotechnology*, 2006, **24**, 255–260.

CHAPTER 10

Wall Paintings: Aspects of Deterioration and Restoration

KARIN PETERSEN

University of Applied Science and Art, Hildesheim/Holzminden/Göttingen, Institute of Restoration and Conservation, Bismarckplatz 10-11, D 31134 Hildesheim, Germany

1 INTRODUCTION

Wall paintings are different from "mobile" paintings executed on canvas or panels, because they represent an integrated part of the building that cannot be moved except by strappo or stacco in times of extreme danger. By such interventions, however, the integral connection is irrevocably lost and cannot be restored. In order to understand the different destructive processes occurring in wall paintings, and any interventions of conservation and restoration required, we have to understand the interactions between the building itself and the wall painting concerned.

This chapter is presented from the point of view of a microbiologist but it will not be confined to the known facts of deterioration by soluble salts and biodeterioration. It will necessarily focus on the interactions between the different causes of damage and the effects of interventions, which may exacerbate the deterioration processes.

2 WALL PAINTINGS AS AN INTEGRAL PART OF BUILDINGS

Wall paintings may be situated in completely different locations, which affect the deterioration process to a large extent. Facade paintings are, for example, subject to direct weathering and the extent to which they are affected by rain, hail, snow or wind depends on the specific construction of the building. Additionally, environmental and air pollution will endanger these paintings in a different way from those inside a building. The protective effect of protruding roofs or drainage systems, such as gutters, becomes obvious when, during poor maintenance of a building, their function is no longer guaranteed and the walls get wet. Moreover, the effect of direct sunlight must not be neglected, since, at different times of the day and the year, these areas may be subject to extreme variations in temperature.

Wall paintings inside a monument are also subject to climatic influences. First, the objects on the outer walls are influenced – depending on the thickness and construction of the wall – more or less directly by the climate outside. This is especially so when the walls are in poor repair, resulting from defective pointing, incomplete outer plaster or sludges. Not only single-hulled walls but also twin-hulled walls may suffer from an input of moisture, damaging the paintings inside the monuments. It is obvious, however, that paintings on inner walls are not usually affected by the climate outside, except perhaps, through strong illumination by sunlight through windows. Walls on upper floors are normally not affected by ascending moisture. However, in buildings without or with defective damp-proof barrier, rising moisture can be a serious danger in the lower parts. The accumulation of earth layers around old churches must be mentioned here. If the soil exceeds the level of the original horizontal damp-proof barrier, this barrier loses its effectiveness. This also takes place in buildings surrounded by rising water level. Depending on the foundation of the building, the moisture not only reaches the outer walls but also penetrates further inside as well.

Paintings on ceilings which are not directly situated under the roof are much less affected by moisture than those situated directly beneath the roof. Unsuitable heating systems may cause deterioration of such paintings by condensation, especially if there is no separation between the roof and the ceiling. More often, water leakages resulting from damaged roofing tiles can quickly cause massive detrimental effects on ceiling paintings. Paintings beneath windows are endangered by water not only when old window frames are leaky, but also when new windows are not fitted correctly or drainage for condensed water is missing. In situations with reduced ventilation, *e.g.* behind altars, large paintings or furniture, humidity may rise too. Last but not least, wall paintings are damaged during the normal use of the room by mechanical damage, heating, dust and other pollution.

In addition to mechanical damage, which may be due to extensive use and may cause a more or less distinct loss of the original paint layer, changes in the substance of a building deserve consideration. Transfers, very often deliberate, or unintended, cause damage by whitewashing or plastering. For centuries, changing styles have caused destruction of wall paintings. New plaster to carry the "modern painting" may have been applied and the original surface may have been damaged by holes made to improve the adhesion of the new plaster. However, neither this damage, nor destruction by earthquakes, collapse, demolition or war will be considered in this chapter.

3   DAMAGING PROCESSES

Deterioration processes, especially caused by soluble salts interacting with moisture and microbial colonisation, will be presented in this chapter, in relation to restoration/consolidation treatments and their effects. Damage will also be considered that has occurred as a later consequence of extended ingress of moisture through damaged parts of the building or the use of unsuitable building materials.

3.1   Deterioration by Soluble Salts

We have to consider that unless the causes of deterioration due to soluble salts have been intensively investigated, the methodology of intervention *in situ* will not be clear. The majority of the wall paintings contain salts like carbonates, chlorides and nitrates of sodium, potassium, calcium, magnesium and ammonia dispersed within the porous material, or concentrated in efflorescences of different aggregates of crystals on the surface or in form of subflorescences below the paint layer. Subflorescences, in particular, cause damage to the wall paintings during repeated cycles of crystallisation and hydration (reversible phase transitions) of salts (from an aqueous solution into the crystalline stage) as a result of climatic changes.

The ions originate from the building materials themselves. Bricks and mortars from coastal areas, may from the beginning, contain the salts present in seawater. Monuments reinforced or isolated with incompatible materials like concrete or Portland cement may suffer from serious problems due to salt deterioration. Air pollution emitted from natural or artificial sources may deposit on the buildings and may then react with the materials. The deposition of sulfur dioxide from combustion, resulting in the formation of black gypsum crusts, is the best-known example. Nitrogen compounds from traffic as well as agricultural sources, aerosols from the sea and chlorides from de-icing salts may contribute to the ionic loading of the aqueous solutions in the porous systems of brick, stone, plaster and the paint layer itself. Furthermore, acids and alkaline solutions applied for cleaning may interact, and result in the formation of additional salts, when reacting with the materials or existing salts.

In addition, several salts are of biogenic origin. In the surroundings of human settlements, chlorides and nitrates accumulate. Microorganisms, like nitrifying bacteria, produce nitrites and nitrates from excrements, waste and even corpses (graveyards), which penetrate building materials by infiltration in rising damp. In addition to this external production of biogenic nitric acid and nitrates, nitrifying bacteria have also been shown to be active in building materials *in situ* and thereby contributing to biodeterioration directly. As well as the nitrogen cycle, microorganisms also contribute to the sulfur cycle and may produce sulfuric acid.

Salt ions are diluted in water and thereby transported into the pore system. Whenever the water evaporates, the ions concentrate and crystallise. The morphology of the crystals is influenced by the conditions at site, especially by other ions present, the microclimate and the internal structures. The disintegration and detrimental effect of subflorescences is mainly due to the pressure produced by the growing crystals as well as hydration.

The process of deterioration happens over a period of time rather than being the result of a single event. Seasonal and diurnal cycles have been reported to be involved (Arnold and Zehnder, 1991). Initiation of hydration may take only minutes rather than hours or days, depending on the specific environmental conditions and material components at site. There are different approaches to prevent these deterioration processes: (i) stabilisation by climate control, (ii) stabilisation by chemical reactions, (iii) extraction of ions and (iv) microbial activity.

Stabilisation by climate control.   The most desirable method to prevent damage induced by the repeated cycles of crystallisation and hydration would probably be environmental control. However, neither the selection nor the maintenance of such an ideal environment is possible if looked at realistically. Predictions of salt crystallisation and hydration from mixed salt solutions are more or less impossible, taking into account all the different parameters that influence the process. Sawdy provides a brilliant overview of the subject and considers relative humidity, temperature, air movement, type and structure of the porous support, salt mixture composition and salt concentration. It is necessary to consider not only consolidation treatments of the plaster or the paint layer, which as such may influence the transition behaviour of the salts, but also the influence of microbial extracellular slimes on the porosity of the system.

The situation becomes even more unpredictable since crystallisation of even pure salt solutions does not occur at a given point of relative humidity. Also the behaviour of the solvent, in this case water, within the porous system of the multi-layered plaster and paint layer system is different to, and much more complicated than, *in vitro* experiments. Therefore, we have to accept that the prediction of crystallisation and hydration cycles of salt mixtures within the porous wall painting system is one of the most difficult tasks one can imagine.

Any kind of monitoring to verify or reject the predictions at site is restricted by the fact that analyses of the efflorescences are necessary to identify precisely which of the possible salts actually do crystallise. Most kinds of monitoring are even restricted to those parts over the paint layer rather than underneath, where crystallisation in lower layers has been taking place. Moreover, attempts to avoid crystallisation of salts in the pore water by enhancing relative humidity of the object has induced enhanced microbial growth, and thereby biodeterioration, in many cases. It should be borne in mind that furniture, painted sculptures, organs and additional decoration nearby may require a different climate to that needed to restrict deterioration by soluble salts in the wall painting.

Attempts at environmental control to stabilise mural paintings were used for the restoration of the paintings in the Tomb of Queen Neferati at Thebes. In this case, the relative humidity outdoors and in the tomb with different numbers of visitors was monitored and compared. The results led to a corresponding restriction in the numbers of visitors allowed into the monument.

Stabilisation by chemical reactions. The so-called Barium method developed in Florence by Ferroni and Dini is well-established and has proved to be a convenient, efficient and durable method of consolidation, after more than 30 years of use. This method depends on the transformation of gypsum. The reaction between ammonium carbonate and gypsum (desulfurication) and application of barium hydroxide produces insoluble barium sulfate, followed by a spontaneous formation of barium carbonate (consolidation). There are well-known limitations due to the presence of copper pigments as well as some organic bindings (like animal glue and vegetable gums), which will not tolerate water from the poultice or undergo cleavage reactions due to strongly alkaline conditions. Therefore, preliminary testing is essential in the case of tempera paintings. Best results are achieved when applied to frescoes and lime paintings.

To overcome the restrictions of the Barium method, ammonium carbonate was, in some cases, replaced by anionic exchange resins that are less aggressive and enable the treatment of tempera even in the presence of copper pigments. It must be added that the reaction of anion exchange resins is restricted to the surface in contact with the resin. Furthermore, the first-stage transformation producing barium sulfate is left out and barium hydroxide is applied directly. This technique is convenient with the many restrictions imposed by preconsolidation by synthetic resins.

Extraction of ions. Methods to reduce salt loadings were, in the beginning, more or less restricted to the application of wet compress materials, *e.g.* cellulose pulp, to the surface of the paint layer. The enhanced humidity leads to hydration and enhanced solubility of ions that are transferred to the surface, and migrate into the pulp material by evaporation of water from the compresses. Repeated applications are necessary and the quantity of ions extracted has to be measured to decide to what extent the extractions should be done. Obviously these extraction techniques can only be successful, if it can be assured that there will be no further transport of ions into the painting, possibly enhanced by the method of attracting the salts into the compress. Hence the paint layer may be endangered by this intervention itself, and even microbial surface colonisation has been observed, even in several situations where none existed before the application of the wet compresses.

To reduce this problem and reduce stress to the paint layer, methods were subsequently developed to include climate control, so that ions are mobilised and transferred to the rear support, in order to reduce salt concentration in the paint layer. This results in a more even salt distribution in deeper parts of the plaster or brick. Ideally, a new layer of plaster can be applied to the back, *e.g.* the supporting brick. The aim of this treatment is to ensure crystallisation within or upon this plaster, but it is difficult to achieve. Investigations nevertheless demonstrated a remarkable reduction of salt ions near the highly-vulnerable paint layer and a more even distribution in deeper parts, which may give more time to search for other solutions. Obviously this technique is restricted to relatively thin supports and cannot be applied in the case of cavity walls.

Other methodologies, such as electro-osmosis, have been reported to be successful on certain objects. In this case, electrodes were introduced in areas where the paint layer was already lost completely to reach beneath the remaining parts of the paintings. Migration of anions and cations of detrimental salts was induced *via* an electrical field. Whereas cations were transported toward earth, anions were attracted by the iron anode. The system is reported to be under observation, with repeated exchange of the iron at the end of its service life. The ions reacting with the iron can be removed from the system and analysed further. A drop in the current of about 300 mA was observed during salt reduction. However, up to now the method has only been applied to a small number of objects so the results will have to be studied carefully.

Microbial activity. Microorganisms are involved in the global sulfur and nitrogen cycles through a wide range of biochemical reactions. In the context of deterioration problems of wall paintings (and stone) caused by repeated crystallisations and hydrations of salts, bacteria involved in desulfurylation and denitrifying processes have been considered for their use in salt-reduction methodologies of bioremediation (see also page 233).

In aerobic respiration, the electrons from carriers at the end of an electron transport chain are finally accepted by molecular oxygen. With oxygen serving as an electron acceptor, as much energy as possible is released from the electron donor. When the environment is depleted of oxygen, alternative electron acceptors, such as sulfate or nitrate, may be reduced. However, in the case of denitrifying bacteria, the aerobic and anaerobic respiration pathways compete, and under normal conditions oxygen as an electron acceptor will succeed (facultative denitrification). In other microorganisms like sulfate-reducing bacteria, the reaction is obligatory and oxygen cannot be used. Hydrogen or organic compounds, like lactate or acetate, function as electron donor, with

H_2S as end-product of the reaction. As mentioned earlier, denitrification is only effective under oxygen depletion. In this situation, nitrate may be reduced to gaseous nitrogen (N_2) by several steps involving nitrite and nitric oxide. Nitrogen is released to the atmosphere. Although this process is disadvantageous for agricultural purposes, it may be of great use for sewage treatment and also in treatment of stone and wall paintings.

Since both processes only take place under oxygen depletion, they have to be carried out under oxygen-free conditions on site using compresses with airtight coverings. Sulfate-reducing bacteria cannot use oxygen and their activity is thereby restricted to the application phase. Denitrifying bacteria, however, may also be active under aerobic conditions and may then contribute to biodeterioration processes. To overcome these problems, attempts have been made to apply the bacteria in an immobilised form. This not only guarantees complete removal of cells after application, but also minimises the problem of biodeterioration afterwards. However, the activity of these immobilised bacteria is restricted to the surface of the objects, where they are in direct contact with the porous materials.

Both techniques are more or less still under investigation in preliminary applications. However, they may be an interesting alternative in the future, if executed with immobilised bacteria, because complete removal of potentially detrimental bacteria by cleaning methods after the treatment is difficult to achieve. Even if biocides were applied for disinfestation purposes, the cell material (biomass) would stay on the surface and in the material pores and provide a nutrient supply for other microorganisms.

3.2 Deterioration by Microorganisms

Factors enhancing microbial colonisation. The essential precondition for microbial colonisation and possible deterioration is the adequate supply of moisture. The required amount may vary considerably for different microorganisms. Even among the commonly encountered species of moulds, there are those that grow in a rather low water environment (*a_w values from 0.75) but also species only growing at a high water activity at ~0.95. In general it is considered that fungi can grow at rather low water levels, while bacteria and algae grow at higher moisture. It must be stressed that moisture is the most essential factor for microbial colonisation and even if all other demands concerning the habitat are met, there will be no microbial colonisation without sufficient moisture. Nevertheless, more preconditions have to be met. For the formation of a microbial community, a range of nutrients have to be available.

---

*a_w is water activity where $a_w = \dfrac{p \text{ (material)}}{p_0 \text{ (pure water)}}$, a measure of available water on a scale 0–1; pure water has an a_w of 1 and the lowest a_w for biological activity is 0.6.

The most important macro-elements can be identified, *e.g.* carbon, nitrogen, sulfur, phosphorus, iron, sodium, potassium and magnesium, which are the main components of the so-called *cell dry-mass*, but other microelements such as manganese, cobalt, copper, nickel, molybdenum, selenium, zinc and vanadium are also required. The carbon sources that can be utilized by different groups of microorganisms can be used to distinguish between autotrophic and heterotrophic organisms. The most important species on wall paintings are autotrophic algae, cyanobacteria and nitrifying bacteria, while the chemoorganotrophic (heterotrophic) bacteria and fungi, which depend on organic carbon compounds, also occur.

In relation to the microbial requirements for humidity, the influence of the climate becomes especially obvious for objects that at some stage have been inaccessible, creating a stable climate. Recently, newspapers reported on microbial problems in a cave in France that had been recently opened to the public. Even if only a limited number of visitors are allowed, it is necessary to consider the amount of humidity that is introduced by people. Therefore, in most cases, wall paintings opened to the public are subject to enhanced humidity, due to both human influences and the effects of the external climate after opening. Only very rarely can investigations of the climate and air contamination be carried out before documentation, investigation and consolidation are started. One of the exceptions to be described is the tomb of Queen Nefetari, in the Queens Valley, Thebes. Caves like Altamira or Lascoux have now been completely closed to tourists and are now only accessible to selected visitors with specific interests.

To enable visitors to observe the paintings and people to work on site, electric lights have to be installed in places which may not have been illuminated for centuries or longer. Light in damp areas inevitably leads to infestation by algae and cyanobacteria, which are photosynthetically-active microbes. Subsequently, chemoorganotrophic bacteria and fungi can grow on the organic compounds synthesized by the light-dependent phototrophs. These successive growths may induce biodeterioration of the paintings. The influence of humidity can be clearly observed in the lower parts of wall paintings inside old churches where, in most cases, algal growth is restricted to areas that are subject to rising damp or other places of enhanced humidity, such as seen in Figure 1 after rainwater infiltration through the vault.

To control the availability of moisture to microbes, ancient techniques of preservation introduced sugar (fruit preservation) or salt (vegetable, fish, meat, hides) to reduce the amount of water available for the organisms. Therefore, in early investigations of biodeterioration of wall paintings, areas with obvious salt problems were expected to be without infestation. However, it was shown that a considerable number of microorganisms isolated from wall paintings are adapted to growth under high salt concentrations. Even pH-values about 11 can support fungal growth. Under such conditions on wall paintings, especially in

Figure 1 *Algae growing on medieval paintings in a vault after infiltration of rainwater, Eilsum, Germany*

areas with highly alkaline pore solutions, such fungi will outcompete other species that are restricted by alkalinity. This should be considered when using cleaning or consolidation materials providing alkaline pH.

The most important deterioration problems on wall paintings, induced or influenced by microorganisms, can be summarised as follows: (i) deterioration due to biofilms, (ii) metabolism of organic bindings as substrates, (iii) alterations of mineral pigments, (iv) excretion of mineral or organic acids, and (v) interaction of microbes and faunal elements.

Deterioration due to biofilms. When considering the biodeterioration of wall paintings, we have to distinguish between alterations associated with living and actively-metabolising cells and those parts of the biomass that are dormant, or even dead. Microbial infestations of wall paintings will contain viable as well as non-viable cells, and may lead to a completely different appearance of the paintings. Algae and cyanobacteria produce chlorophyll to form green layers upon the surface. Fungi are frequently pigmented by brown melanins, and several bacteria express red or yellow carotenoids. However, even cells without bio-pigments change the appearance by covering the painting, which might be looked upon as being merely an aesthetical problem. Even if there was no influence on the paintings, as was suggested for some bacteria causing red discolourations on wall paintings, one has to consider possible effects on humans, as will be discussed later.

Microbial growth frequently leads to the production of so-called biofilms from extracellular polymeric substances (EPS) surrounding the cells from which they have been produced. Biofilms, including the cells themselves as well as surrounding EPS, may influence or induce deterioration, *e.g.* by swelling and shrinking during climatic changes (Figure 2) that induce condensation or evaporation processes and thereby may also influence the state of soluble salts. The biofilm as a superficial layer reacts in a detrimental way comparable with poor surface consolidation, *e.g.* casein which has not been applied in an appropriate way. Additionally, the application of surface consolidants, even cleaning, procedures may be influenced and limited by the presence of biofilms.

Frequently, however, growth is not restricted to the surface. Even those microorganisms that are dependent on photosynthesis, based on carbon dioxide fixation in the light, may develop successfully just beneath the surface, *e.g.* in renderings and fillings. Fungi often form more or less tissue-like mycelia beneath the paint layer and have been detected in areas where casein was injected as binder for consolidation treatment of plaster (see Figure 3).

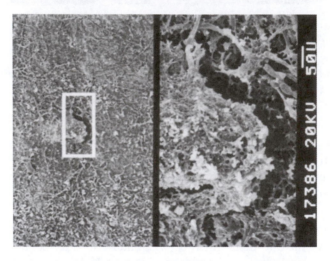

Figure 2 *SEM of biofilm on wall paintings, Kloster Altenburg, Austria*

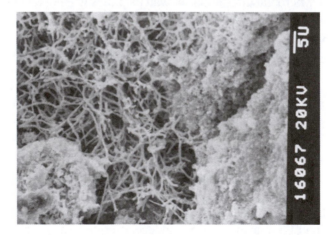

Figure 3 *SEM of microbial colonisation beneath plaster, Eilsum, Germany*

The interfaces between different materials of the multilayered wall painting system may be heavily attacked by microbial growth. Microbial biomass developing beneath the paint layer will at least partly destroy the connection between paint layer and the plaster below and thereby induce flaking, especially when larger parts are attacked and when climatic changes take place. Similar situations may occur beneath applications made over paintings, resulting in uncontrolled exfoliation. The problem is worse in cases where microbial subsurface growth is accompanied by salt crystallisation in the same layer (Figures 4 and 5), as frequently observed.

Metabolism of organic bindings as substrate. It is well known that natural organic bindings, like animal or plant glue, lose their binding effect when

Figure 4 *Flaking of paintlayer, sample from Gnoiden, Germany*

Figure 5 *SEM of subsurface layer shows microbial growth and salt efflorescences*

they are contaminated or when they have been in use for an extended time, and then consolidation is required. Microbial contamination of organic bindings leads to enzymatic breakdown of macromolecules like proteins, fatty acids or carbohydrates because microorganisms are unable to absorb these large macromolecules. They produce exoenzymes to break them down into smaller monomers, which can then be absorbed by cells as carbon source. The polymers then disintegrate and can no longer function as consolidants. In any paintings not executed as fresco, or at least as lime-painting, organic bindings guarantee the consolidation of paints. During microbial infestation, these organic bindings will be decomposed and lose their consolidation effect. Metabolism of organic bindings is not only restricted to natural organic bindings but also affects synthetic polymers applied for consolidation purposes.

Alteration of mineral pigments. Besides the production of bio-pigments, microorganisms may also interfere with the mineral pigments of the painting composition itself and thereby completely change its appearance. Sulfur-reducing bacteria may produce black sulfides from lead white. A summary of pigments transformed by microbes is given by Lyalikova and Petrushkova. Lead oxides, frequently suspected of being transformed by microbial activity, however, could not be detected in the darkened ceiling of Sala Terrena in Weißenfels Castle, Pommersfelden, although microbial infestation could be shown. In laboratory experiments, production of acetic acid by fungi isolated from wall paintings resulted in the formation of green copper acetate from blue azurite. The reaction *in situ*, however, is still in question. Other pigments like manganese or iron may be oxidized by fungi isolated from building materials and thereby change their colour. As mentioned above, microorganisms require certain metals for growth. If these elements are limited, the microorganisms are able to extract the minerals from the pigments that the paintings are composed of, and to incorporate these into biomass.

Excretion of mineral and organic acids. Bacteria growing by nitrification and sulfur oxidation excrete mineral acids into their surroundings while gaining energy from these reactions. Sulfuric acid and nitric acid are strong mineral acids, which induce deterioration of components of the painting, plaster or stone, particularly lime, but also organic bindings in the objects.

Excretion of organic acids may be connected to the turnover of specific nutrients and to mineral limitation. Acids frequently excreted by microorganisms, especially when growing on carbohydrates, are, among others, oxalic acid, citric acid and acetic acid, which damage wall paintings by the same mechanisms described above. In some cases, however, the excretion of oxalic acid into calcareous materials may result in the formation of resistant oxalates. After oxalates had been detected on many ancient marble statues, it led to the suggestion that microbial activity might have been responsible. This will be discussed later on, in the context of consolidation treatment. Discolouration of mineral pigments due to excretion of organic acids has already been considered.

Interaction of microbes and faunal elements. Besides the stimulatory and competing interactions between different microorganisms, coexistence of microbes with faunal elements like mites, springtails and even mosquitos can be observed by means of microscopy, especially scanning electron microscopy. Faecal pellets, eggs, silk and exuviae sampled from wall paintings have been shown to be usually infected by fungi and bacteria. Despite the well-known symbiotic interactions between arthropods and fungi within guts, little or no investigations have been done to assess other possible impacts.

As for the animals, they may feed on microbes directly or may use microbial enzymes for digestion purposes. The microbes, however, may profit by the fact that they can be transported over relatively long distances and thereby colonise large parts of the paintings, as long as the animals are mobile and carry the microbial cells on their body. Some of them, however, may also be transferred after passing through the insect digestion system.

For example, flies may already be infected by fungi while still alive and mobile; dead individuals with spore-bearing conidiophores growing from their body have been observed. Some mosquitos sampled from wall paintings (Kodersdorf crypt of a parish church, dating back to the eighteenth century) immediately after dying did not prove to be infested (see Figure 6) but later, however, the exuviae were completely overgrown (see Figure 7), especially by the fungus *Verticillium*.

Although some species of the genus *Verticillium* have been described that grow on chitinous substrates (important components of the insect exoskeleton),

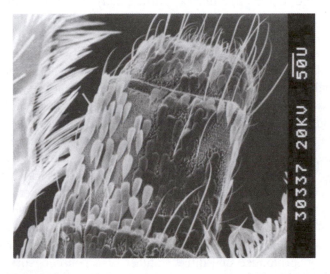

Figure 6 *SEM of sample from Kodersdorf, Germany, mosquito body without infestation*

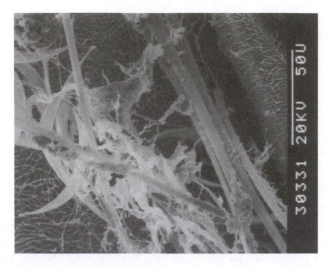

Figure 7 *SEM of sample from Kodersdorf, Germany, mosquito body now heavily infested by fungi (seen as filaments)*

this effect could not be shown for isolates cultivated from the samples. However, they were able to grow on blood agar in laboratory experiments. Blood proteins and fats are provided by the mosquitos. Excrements (faecal pellets) from arthropods, as well as bats and mice, provide large amounts of different organic nutrients and have been shown to be heavily infested by microorganisms, especially fungi. In addition, silk protein and eggs are also available as growth substrates.

Grazing might also provide a kind of stimulatory effect on microbial growth and deterioration after mechanical disturbances of the biofilm. However, we have to consider that there might be a direct detrimental input of arthropods on mural paintings as well, since grazing is not restricted to the biofilm but may destroy the painting or plaster directly.

As in other terrestrial habitats, microbes and especially fungi may act as symbionts, parasites and decomposers in the ecosystem of the wall painting. The possible ecological connections can be summarised as follows: (i) decomposition of accumulated organic matter, including dead insect bodies mainly by saprophytic fungi, (ii) parasitism on living arthropods resulting in their death and thereby making organic matter available for saprophytes, and (iii) transport of surviving microbes during grazing for nutrition by arthropods.

It is only very recently that other aspects, such as health risks to human beings, have been considered. Possible allergic reactions of atopic people in indoor environments, which are heavily infected with fungi, are known from other fields of interest but have also been discussed in the context of the initial opening of a Pyramid ("Der Fluch des Pharaos" – the curse of the Pharaohs). The intensive, prolonged and direct contact of a restorer working on site, especially while dry cleaning or uncovering large areas of the paintings, may pose a potential risk. Many fungi on wall paintings produce an enormous amount of allergens, associated with dead cells according to their growing conditions, but also toxins, such as cell wall toxins. Toxins from bacteria and mycotoxins from fungi, have been obtained from some of the species detected on wall paintings. Up to now there have been few investigations of mycotoxin production at such sites. Recently, we could detect ochratoxin in dust samples from St. Martin (Kreiensen-Greene), which has wall paintings dating back to 1575 that were covered by lime wash in 1716 and uncovered in 1977. The paintings are heavily overgrown by the fungus *Aspergillus glaucus* (*Eurotium herbariorum*) on an area of about 50 m^2 of original surface. About 30% of *Eurotium herbariorum* isolates are known to be possible producers of ochratoxin depending on how they are grown in the laboratory. In experiments we could show that some isolates from other objects were also capable of toxin production. To complete the investigations, dust samples from the church were analyzed for ochratoxin and also turned out to contain this mycotoxin (unpublished).

An even more dangerous situation arises in people with defective immune systems with thermotolerant fungi that are able to grow at temperatures as high as 37°C (body temperature). These fungi are normally eliminated by the human immune system, but if acquired by people with immune insufficiency, they may be lethal. This topic is more or less completely unexplored; this includes the extent to which these species are distributed at different sites, their ability to produce mycotoxins under the prevailing climatic conditions and on the substrates available. Therefore, this is likely to be one of the most urgent questions to be worked out over the next few years. Until then, risk should be avoided and conservation jobs with this type of high risk potential should only be carried out with specific safety precautions.

4   CLEANING AND CONSOLIDATION

Aspects of cleaning, as well as consolidation, have already been discussed when dealing with the "barium method" of extraction and consolidation. So in this section, cleaning in the context of salt extraction will not be referred to anymore. However, there are other materials that become part of the paintings over centuries of intensive use, as well as previous conservation treatments. Many of these materials were applied as a consolidation treatment. This section will therefore consider different consolidation methods before discussing how to reduce their sometimes detrimental effects.

During consolidation of plaster and rendering, problems arising may be the same as that of surface consolidation. However, to see these changes beneath the paint layer in the early stages may be much more difficult. Modern materials like dispersed white lime, contain minor organic ingredients to keep them dispersed or to stabilise the product, and these may serve as organic substrates for microbial infestation. In these circumstances, one then has to consider the risk of biodeterioration.

4.1 Consolidation

Consolidation generally can be divided into two types of methods: inorganic or organic consolidants. The latter may be natural organic polymers or even synthetic polymers. Both types of treatment may cause longer-term drastic alterations although acceptable results may have been obtained directly after application.

Consolidation by inorganic materials.   Up to now, a mineral treatment such as the application of barium hydroxide (described earlier) seems to be the only method that does not turn out to be detrimental in the end, as long as it is applied carefully and under the right conditions.

Application of "Wasserglas" (sodium silicate) as a consolidant for wall paintings came into use at the beginning of the nineteenth century. It was used for consolidation of stone monuments, plaster, lime renderings and also as a binding for silicate paints. Soon after, however, failure of the application became common and the formation of crusts was often seen, leading to enhanced deterioration by flaking. When used in combination with Portland cement, tremendous amounts of sodium carbonate and potassium carbonate are formed. Although potassium carbonate is hygroscopic in itself, and should not crystallise in solution, it may form dark spots on the paintings. Moreover, in paintings with autochthonous salts, the transformation to less hygroscopic salts may be induced and this lead to efflorescences.

In cases of high humidity and known infestation by heterotrophic microorganisms, consolidation treatments based on mineral, rather than organic components, are more appropriate. Therefore consolidation by silicate-based materials was introduced (methyl or ethyl silicate). This led to human health risks. Besides the known risk of silicate formation on the cornea, that may lead to blindness due to methyl silicate, both forms may irritate the respiration system. Moreover, laboratory investigations as well as observations at treated sites, showed an unexpected intensive growth of fungi on ethyl silicate (Figure 8). This is difficult to explain since the catalysts within the products belong mainly to the group of organo-tin compounds, known to be very potent biocides. Until now, the cause of these findings cannot be explained.

Consolidation by organic materials.   More commonly, organic polymers have been applied for consolidation of paint layers. Natural polymers were most frequently applied in Europe, including animal glue, casein, egg yolk

Figure 8 *Infested sample after consolidation with ethyl silicate, Clemenswerth Castle, Soegel, Germany*

and, especially in England, bees-wax, while linseed oil was basically used to make the paintings more brilliant.

After the application of surface consolidants that reduce porosity, problems arise in the presence of autochthonous salts (especially with changes of the environmental climate). Additionally, the consolidants themselves react to climatic changes by shrinking and swelling. This results in flaking of the consolidant, as well as the pigments, after repeated cycles. Therefore cleaning, or at least reduction of the consolidation layer, must be undertaken to preserve the paintings that have undergone this kind of consolidation.

In the second half of the twentieth century, natural polymers were subsequently replaced by synthetic polymers or derivatives of natural polymers, *e.g.* based on cellulose, like methyl celluloses or hydroypropyl celluloses. The most frequently applied synthetic materials range from polyvinyl alcohol and polyvinyl acetate to acrylics. Even nylon was used. Deterioration by this type of consolidant was enhanced at sites with autochthonous salt problems, and this in turn enhanced the activity of bacteria and fungi. It could be clearly demonstrated that microorganisms could unexpectedly metabolise synthetic polymers. All synthetic polymers investigated in laboratory experiments turned out to be biodegradable and supported the growth of microorganisms. It was not only the polymer itself but also numerous additives in the product that supported growth. Microbial infestation of paintings with synthetic polymer coatings can frequently be observed, especially by scanning electron microscopy. However, although one can prove the use of the polymer as substrate in laboratory investigations, this does not necessarily mean that the microbes *in situ* can use the synthetic polymer as organic carbon source. It is

Figure 9 *SEM of microorganisms growing beneath Klucel surface consolidation, Salzhemmendorf, Germany*

more likely that natural organic materials like casein or glue may function as a carbon source. Nevertheless, excreted microbial intermediates like organic acids may attack the synthetic polymers and especially synthetics applied as a coating. This may result in some kind of greenhouse effect by reducing desiccation, when applied over an existing microbial infestation (see Figure 9).

Application of wax and various wax resin mixtures was predominantly employed in England until the middle of the last century as a result of misunderstanding of the ancient Roman techniques. The application caused detrimental results. After it was agreed that wax as a consolidant was not appropriate, subsequently techniques of de-waxing had to be devised, which may, however, only reduce the amount of wax in the painting, not remove it completely. Moreover all those removal techniques applied until now involve use of organic solvents, with enormous risks both to paintings and conservators.

Consolidation by alternative approaches. The latest approach is the application of oxalic acid to wall paintings, as well as stone monuments. This method is still under investigation in Italy. After intensive laboratory investigations, some *in situ* applications on stone monuments were executed. The proposed advantage is the controlled transformation of the uppermost layer of calcium carbonate into oxalate, which is much more resistant to any kind of aggressive environmental attacks. This can be observed on a large number of architectural surfaces with compositional oxalate films. As compared to calcium carbonate, the oxalate is much more resistant to acids. At neutral pH the solubility of calcium oxalate is about two orders of magnitude smaller than that of calcium carbonate. It still has a very low solubility down to pH values of about 2–3. However, the treatment by oxalic acid is completely irreversible and, as soon as micro-fissures or other lesions occur, water and salts from the atmosphere start penetrating behind the coating and will subsequently lead to efflorescences. Therefore an artificially-induced oxalate layer should permit some porosity. First attempts to apply microorganisms at site to induce excretion of oxalic acid, and then transformation, were restricted, because it would have been difficult to restrict microbial activity to a particular level – a problem that has been discussed before.

Calcification by bacteria. Lime-water application seems not to guarantee sufficient consolidation. Precipitation of calcite by microorganisms is known to occur in freshwater and marine environments, as well as in soil, and this led to trials of bioremediation on heritage surfaces. Very recently, approaches using calcification due to bacterial activity have even been carried out particularly on stone and architectural surfaces (see page 236). Remediation by several bacteria was shown to result in the formation of newly developed calcite precipitations. For this, chemoorganotrophic bacteria, as well as nutrients, were supplied to the surfaces. Up to now the role of bacteria is not completely understood, however, studies on marine bacteria revealed three main mechanisms: (i) calcium-binding membrane proteins, (ii) ionophores, and (iii) extracellular materials on cell surface.

These bioremediation treatments are based on the assumption that bacteria will not cause biodeterioration at some point in the future after the applied nutrients have been exhausted. If the new precipitated calcite completely encloses the cells, this might be true. If this is not true, there is no reason why they should not live on other nutrients available from the environment.

Laboratory investigations resulted in bio-induced calcification from both dead cell material and organic matrix macromolecules from the shells of molluscs. Different results were obtained by accident after lime impregnation was done on site in Denmark. Calcite precipitation on fungal hyphae was demonstrated after impregnation, which resulted in at least a temporary consolidation. However, fungi have to be considered as potential detrimental microorganisms, especially if calcification does not result in embedding of the cells, so inhibiting exchange with the environment and resulting in cell death. Interestingly enough, carbon dioxide release by cell respiration may also contribute to calcification but this has not been considered.

To summarise, bioremediated calcite precipitation as a consolidation treatment for wall paintings has not yet reached the stage of development to be recommended without more intensive tests.

It is clear that a consolidation treatment for wall paintings suitable for all situations cannot be expected in the near future. The treatment will be determined by the amount of consolidation necessary, the material and the form of application possible at the specific site. It is also necessary to take climate into consideration, as well as salt and microbial deterioration processes that may be occurring in the painting.

4.2   Cleaning Methodologies

Cleaning methods are subject to fashion. It is quite clear that any kind of dust upon the surface should be removed from the paintings, not only for aesthetic reasons, but because it attracts humidity and may serve as a nutrient source for microorganisms.

There are many different approaches, for example, application of cattle faeces to remove soot (which is one of the most predominant surface pollutions on paintings), application of acids or application of organic solvents. Even bread or dough was commonly used in previous times and it is still a practice in Scandinavia at the present.

Cleaning by aqueous and organic solvents.   Mineral acids such as fluoric and hypochloric acid, as well as organic acids like formic and acetic acids, have been used for cleaning purposes, especially after mural paintings have been uncovered from the superficial layers of whitewash, with detrimental effects. When applied in liquid form or in poultices, it is not only the direct attack on carbonate that occurs. As described for alkaline compounds and ionic tensides, ions remain in the paintings and form salts that cause new problems. Although this is known, it is often forgotten that neutralisation of such treatments entails the formation of salts. Application of ion exchange resins has already been mentioned.

Before organic solvents are applied, the chemical nature of the materials to be removed should be intensively investigated. Only then can the solvents appropriate for application at site be tested. This is especially important, since most of the popular solvents bear enormous health risks. The risk may be reduced by the application of solvents in the form of gels as introduced by Wolbers.

Cleaning by biochemical methodologies.   Recently, reports on the application of bacteria for cases of flaking mural paintings caused by consolidation with glue, have been published. Attempts were made to solve the problem by microbial activity. It was suggested that the bacteria applied are washed away after treatment. However, wall paintings provide a more or less rough surface, so that it is impossible to remove all bacteria completely. Consequently, the application of biocides has to be considered to prevent the problems arising from the microorganisms themselves. Moreover, how can it be guaranteed that amino acids, resulting from the extracellular enzymatic cleavage of protein, do not remain in the paintings as a readily-available substrate for microbes? And how is it possible to guarantee that the bacteria only reduce surface consolidants, and do not attack the original binding in cases of paintings not executed a fresco? There are still concerns about this approach.

Figure 10 *SEM of impregnation/consolidation at Schongauer paintings, Berisach, Germany*

During the last decade of the twentieth century, enzymes came into use for reduction of such coatings, which had resulted in tension and later led to flaking. First attempts were, for example, done on the Schongauer paintings at Breisach (Germany) as an "impregnation lacquer"; applied in the 1950s, they resulted in flaking and loss of the paint-layer (Figure 10). First scientific examinations revealed a proteinaceous surface coating. The application of proteolytic enzymes in poultices, however, did not give convincing results. After intensive investigations, it turned out that the coating was a mixture of proteins and oily components, which, of course, were not disintegrated by proteolytic enzymes.

Similar studies were part of a national research project in Lower Saxonia funded by Deutsche Bundesstiftung Umwelt (AZ. 10631). To overcome the problems mentioned above, and to avoid the application of organic solvents (and health risks arising from them), a method was developed to reduce inappropriate casein surface consolidations and allow the extraction of the free amino acids during the application. No additional substrates were left in the paintings and the effect of reduction can be easily controlled during the application at site by non-destructive monitoring. This guarantees that paintings containing casein or glue as original binding media, can be treated to remove surface consolidants without damaging the original binding. In addition, the methods used were based on immobilized enzymes, which guarantees that there will be no enzymatic activity left on the painting after the application, since the bioactive molecules will be completely removed from the wall, together with the membrane that was coated with the enzymes. This may also be the only way to make good use of controlled microbial activity during cleaning and consolidation of wall paintings. It should be mentioned,

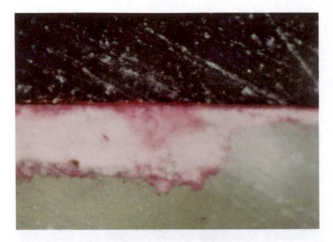

Figure 11 *Protein staining of sample before enzymatic treatment*

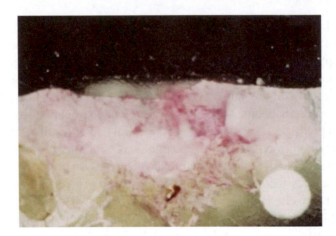

Figure 12 *Protein staining of sample after enzymatic treatment*

that phosphorus can be a growth-limiting factor in ecosystems and therefore one should avoid buffer systems based on phosphorus.

Laboratory investigations on artificial samples demonstrated a reduction of the coating using scanning electron microscopy as well as protein staining (Figures 11 and 12). Application to other types of consolidants should be possible. Membranes with appropriate immobilized enzymes have to be designed according to the different materials applied for consolidation. The system developed in Lower Saxonia is now being adapted to be applied throughout Europe on several objects with deterioration problems caused by casein coatings.

In other recent work, Heiling compared the suitability of bread or sponges (Wishab) for cleaning wall paintings. The results turned out to be completely unexpected. Although most of the microbial isolates from the paintings under investigation were able to grow on starch (from bread), enhanced microbial infestation at the treated sites could not be observed. Although starch could be shown to remain upon the surface in some cases, microbial infestation using starch at site was not seen by scanning electron microscopy, even at sites which were infested in untreated areas. It could be shown that the amount and form of copper in the dough influenced the rate of microbial turnover in laboratory experiments. Moreover, extracts of "Wish up" were demonstrated to serve as nutrient source for the microbes frequently isolated from wall paintings. In relation to water repellent capacity, "Wish up" resulted in a partly enhanced repellence as compared to areas treated with bread. Obviously, these results are being confirmed by a more intensive survey.

5 CONCLUSIONS

It is clear that there are many different factors that induce, or at least influence, deterioration processes in wall paintings, most of them somehow linked with humidity. Attempts to maintain humidity control are difficult and restricted by the fact, that different parts of the room, like furniture and organs, may demand a different climate to that of the painting. For the painting, deterioration processes due to salt crystallisation attracts higher moisture, while, biodeterioration is enhanced by humidity. Consolidation treatments may also induce other detrimental processes that, on the one hand, may result in flaking due to climatic changes, and on the other, may serve as a carbon source for microorganisms.

Methodologies and materials that have been used for a long time are better understood than those that are recently developed. Therefore new approaches should only be tested in controlled circumstances and afterwards be carefully monitored for an appropriate time, before the methodology can be accepted for safeguarding the paintings. Wall paintings should not become a testing ground or field of experimentation for factories producing materials for building maintenance. Moreover the different parameters that lead to deterioration at different places will never allow one mode of intervention, since every situation demands intensive investigations of the detrimental processes and an approach that is specifically adapted to the situation at site, including the state of the building itself, climatic situation, the mode of specific use of the objects, any salt problems and possibly the biodeterioration taking place.

To summarise the problems, one has to consider that any material introduced into wall paintings for consolidation, cleaning or any other purpose may react with ions already present at site and cause new, and even worse, problems.

264   *Chapter 10*

REFERENCES AND FURTHER READING

H. Arai, *Sci. Conserv.*, 1988, **27**, 19.

A. Arnold and K. Zehnder, in *The Conservation of Wall Paintings*, S. Cather (ed), Tien Wah Press Ltd, Singapore, 1991, 103.

A. Ballentyne, in *Les Anciennes Restaurations en Peinture Murale*, M. Steffanaggi (ed), SFIIC, Champ-sur-Marne, 1993, 143.

F.S. Barbaresi, G. Mastromei and B. Perito, in *Molecular Biology and Cultural Heritage*, C. Saiz-Jimenez (ed), Swets & Zeitlinger, Lisse, 2003, 209.

S. Beutel, K. Klein, G. Knobbe, P. Königfeld, K. Petersen, R. Ulber and T. Scheper, *Biotechnol. Bioeng.*, 2002, **80**(1), 13.

I. Brajer and N. Karlsbeck, *Stud. Conserv.*, 1999, **44**, 145.

D. Dini, in *Les Anciennes Restaurations en Peinture Murale*, M. Steffanaggi (ed), SFIIC, Champ-sur-Marne, 1993, 137.

P. Friese, *Arbeitshefte des Bayerischen Landesamtes für Denkmalpflege*, 2001, **104**, 141.

A.A. Gorbushina and K. Petersen, *Int. Biodeter. Biodegr.*, 2000, **46**, 277.

R. Gowing, in *Preserving the Painted Past*, R. Gowing and A. Heritage (ed), James and James (Science Publishers), London, 2003, 85.

I. Hammer, in *The Conservation of Wall Paintings*, S. Cather (ed), Tien Wah Press Ltd, Singapore, 1991, 13.

K. Heiling, Diploma Thesis, University of Applied Sciences and Arts, Hildesheim, F. R. G., 2003.

C. Heyn, K. Petersen and W.E. Krumbein, *Kunsttechnologie und Konservierung*, 1996, **10**, 87.

R.J. Koestler, in *Of Microbes and Art*, O. Cifierri, P. Tiano and G. Mastromei (eds), Kluwer Academic/Plenum Publishers, New York, Boston, Dordrecht, London, Moscow, 2000, 153.

W.E. Krumbein, V. Schostak and K. Petersen, *Kieler Meeresforschung*, 1991, **8**, 173.

L. Laiz, D. Recio, B. Hermosin and C. Saiz-Jimenez, in *Of Microbes and Art*, O. Cifierri, P. Tiano and G. Mastromei (eds), Kluwer Academic/Plenum Publishers, New York, Boston, Dordrecht, London, Moscow, 2000, 77.

P.K. Larsen, Ph.D. Thesis, Department of Structural Engineering and Materials, Technical University of Denmark, Lyngby, 1999.

N.N. Lyalikova and Y.P. Petrushkova, *Geomicrobiol. J.*, 1991, **9**, 91.

S. Maekawa, in *Art and Eternity – The Nefetari Wall Paintings Conservation Project*, M.A. Corzo and M. Afshar (eds), The J. P. Getty Trust, 1993, 105.

M. Matteini, in *The Conservation of Wall Paintings*, S. Cather (ed), Tien Wah Press Ltd, Singapore, 1991, 137.

M. Matteini, in *Preserving the Painted Past*, R. Gowing and A. Heritage (eds), James and James (Science Publishers), London, 2003, 85.

G. Orial, S. Castanier, G. le Metayer and J.F. Loubière, in *Biodeterioration of Cultural Property*, K. Toishi, H. Arai, T. Kenjo and K. Yamano (eds), Toyo Agency Inc., Tokyo, 1993, 98.

K. Petersen, in *Die Restaurierung der Restaurierung*, M. Exner, U. Schädler-Saub and K.M. Lipp (eds), Verlag, München, 2002, 259.

K. Petersen and I. Hammer, in *Biodeterioration of Cultural Property*, K. Toishi, H. Arai, T. Kenjo and K. Yamano (eds), Toyo Agency Inc., Tokyo, 1993, 263.

K. Petersen, U. Mohr, T. Oltmanns and Y. Yun, *Arbeitshefte des Bayerischen Landesamtes für Denkmalpflege*, 2001, **104**, 221.

G. Ranalli, C. Belli, C. Baracchini, G. Caponi, P. Pacini, E. Zanardini and C. Sorlini, in *Molecular Biology and Cultural Heritage*, C. Saiz-Jimenez (ed), Swets & Zeitlinger, Lisse, 2003, 243.

G. Ranalli, M. Matteini, I. Tosini, E. Zanardini and C. Sorlini, in *Of Microbes and Art*, O. Cifierri, P. Tiano and G. Mastromei (eds), Kluwer Academic/ Plenum Publishers, New York, Boston, Dordrecht, London, Moscow, 2000, 231.

A. Sawdy, in *Preserving the Painted Past*, R. Gowing and A. Heritage (eds), James and James (Science Publishers), London, 2003.

P. Tiano, in *Molecular Biology and Cultural Heritage*, C. Saiz-Jimenez (ed), Swets & Zeitlinger, Lisse, 2003, 201.

P. Tiano, L. Biagiotti and S. Bracci, in *Of Microbes and Art*, O. Cifierri, P. Tiano and G. Mastromei (eds), Kluwer Academic/Plenum Publishers, New York, Boston, Dordrecht, London, Moscow, 2000, 169.

R. Wolbers, *Cleaning Painted Surfaces: Aqueous Methods*, Archetype Books, London, 1999.

S. Wörtz, Diploma Theses, University of Applied Sciences and Art, Hildesheim, F. R. G., 2001

CHAPTER 11

Conservation of Ancient Timbers from the Sea

MARK JONES[1] AND ROD EATON[2]

[1] The *Mary Rose* Trust, HM Naval Base, Portsmouth, Hampshire, PO1 3LX, UK
[2] The School of Biological Sciences, University of Portsmouth, Hampshire, PO1 2DY, UK

1   INTRODUCTION

This chapter deals with damage to maritime archaeological timbers caused by the activities of microorganisms, wood-boring animals and insects in wood exposed to and removed from the marine environment. It is important to recognise from the outset that the deterioration of wooden shipwrecks, artefacts, structural timbers and even submerged wooden monuments does not cease when removed from the sea. Inadequate storage of the material prior to conservation will often allow the processes of deterioration to continue. Our intention here is to review the biology of the deteriogenic organisms of ancient maritime timbers, to highlight the hazard they pose, particularly to shipwreck timbers, and to examine the strategies and methods currently adopted to conserve timbers excavated from the seabed.

2   THE BIODETERIOGENS

2.1   Marine Wood-Boring Animals

Wood-boring molluscs and crustaceans are the most destructive agents of timber submerged in seawater. Indeed the marine environment is recognised as the most hazardous situation for exposed timber. The rate at which these organisms attack is dependent primarily on water temperature and the natural durability of the wood species, notably the heartwood which has greater natural resistance than sapwood.

The wood-boring molluscs comprise the teredinids or shipworms (family Teredinidae), and the pholads or piddocks (family Pholadidae). Both groups are bivalves and burrow into wood through the rasping action of the two valves or shells at the anterior end of the animal. Although the teredinid family is the larger of the two groups and is distributed world wide, the occurrence of individual species can be restricted within a range of water temperatures. In contrast, the pholads are found mainly in the warmer waters of the tropics and subtropics, although some members do occur in cold-water situations, either at depth or in higher latitudes. Most of the wood-boring crustaceans are members of the Isopoda – the family Limnoridae or gribble and the family Sphaeormatidae or pill bugs. A third group, the family Cheluridae are members of the Amphipoda. The wood-boring crustaceans have segmented bodies and are able to move over the surface of wood, unlike the molluscs that remain in their burrows for life. Gribble attack of wood is superficial and the animals excavate a network of narrow galleries on the wood surface to produce an hourglass shape in the inter-tidal portion of vertical structural timbers. The chelurids excavate wider galleries, often enlarging those formed by limnoriids, and are known to ingest the faecal pellets of limnoriids. The sphaeromatids are larger in size than the limnoriids or chelurids and usually burrow into wood by tunnelling across the grain, sometimes producing a honeycomb of tunnels in softened timber. Species of limnoriid, and to a great extent the chelurids, have a world-wide distribution from cold temperate to tropical zones, whereas the sphaeromatids occur in tropical, especially brackish waters. In short, the main hazard to archaeological timbers in cooler temperate waters exists from shipworm, gribble and chelurids, while in warm temperate–tropical zones, pholads and sphaeromatids pose an aggressive additional threat. Only a marked reduction in the salinity of major bodies of water, such as the northern part of the Baltic Sea, offers any natural defence against wood-boring animals in non-polar regions.

Teredinids. The family Teredinidae is made up of many genera. When mature, the animals have a soft, worm-like body that in extreme cases can be up to 2 m long. The animals excavate tunnels within the wood by the grinding action of the anterior valves of the worm (Figure 1) and a calcareous deposit is laid down over the surface of the tunnel. This calcareous lining is clearly visible in X-radiographs of timber specimens infested with shipworm and is a useful way of assessing the extent of internal damage to a piece of wood which may otherwise appear superficially intact (see Figure 2). The posterior end of the worm supports two siphons that protrude through a narrow hole (1–2 mm in diameter) at the wood surface. The incurrent siphon draws in oxygenated seawater to support respiration by the animal and feeding on particulate organic matter and microorganisms, while the excurrent siphon releases waste products and reproductive units.

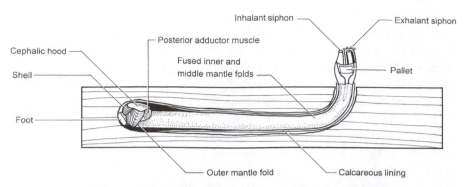

Figure 1 *Diagram of an entire organism (Teredo) showing relative position of shell and siphons*
(After Turner, 1971)

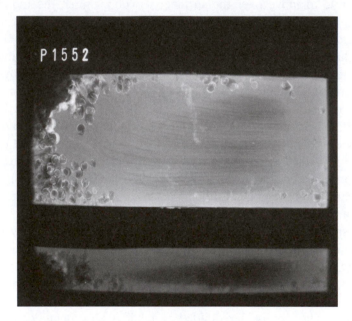

Figure 2 *X-ray radiograph of timber infested by Teredo sp.*

Infection of fresh wood occurs after completion of the reproductive process followed by settlement and attachment of larvae to the wood surface. There is some dispute about the need for larvae to crawl to regions of wood already softened by superficial microbial decay, or whether they can burrow directly into intact wood. In either case, the point on the surface of wood where a larva begins burrowing is the animal's only contact with the external environment throughout its life. During periods of stress, *e.g.* temporary anaerobiosis or exposure above the waterline, this fine aperture is plugged with a calcareous pallet positioned close to the syphons. The morphology of the pallet varies from spade-shaped to a series of ornamented cups and is an important diagnostic feature for identifying species of shipworm. In addition, the delicate nature of the pallets of some species demands extreme care when extracting animals from wood for identification.

Experiments to measure the rate of growth of shipworms in wood have been carried out by repeated X-ray analysis. Rates vary according to the exposure conditions and the wood species tested, but values of 18 mm per month by *Lyrodus pedicellatus* and 74 mm per month by *Bankia setacea* have been recorded. However, where high levels of infestation occur, overcrowding in the wood results in slower rates of tunnelling and smaller diameter worms. Actively burrowing worms do not intrude into neighbouring tunnels, nor do they emerge through the wood surface, even though the wood may be totally destroyed internally. The wood fragments ingested by the worms may be broken down in the gut by cellulolytic enzymes produced by specialised cells of the digestive diverticulae. However, there is now strong evidence to show that microbial endosymbionts play an important part in the digestive process. Cellulolytic, nitrogen-fixing bacteria in the gills of shipworms are also thought to contribute to the nutrition of animals that cannot be wholly dependent on naturally low-nitrogen woody substrates. This aspect of shipworm biology will be dealt with later.

Pholads. The family Pholadidae comprises bivalve molluscs that are wood-borers, but also includes genera that bore into soft rock, firm mud and even concrete. Although wood-boring pholads can be very destructive, the general view has been that they do not digest the wood that they excavate and their burrows simply afford protection allowing the animals to filter feed on plankton. However, recent studies indicate that this may not be totally correct since bacterial endosymbionts similar to those associated with teredinids have been observed within the gills of members of the *Xylophaga*. The burrows of mature pholads can be 5–10 mm in diameter and range in depth from 3 to 8 times the length of the animal (Figure 3).

There is no calcareous lining and apart from members of the Xylophaginae, the majority of pholads lack the soft worm-like body that is so characteristic of teredinids. However, the two calcareous valves and accessory calcareous plates between the valves can be recognised on X-radiographs of damaged wood. Indeed, the morphology of the valves and plates plus the siphons is the main feature to aid identification. The pholads are typically found in warm water situations, although species of *Xylophaga* have been recorded in cold, as well as tropical, latitudes. They also show some tolerance to environmental extremes and can be found in brackish estuarine waters and shallow waters where high temperatures occur.

Gribble. Members of the Limnoriidae are dominated by the genus *Limnoria* and 51 species have been recognised recently. Two other genera of

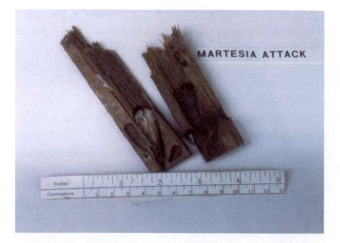

Figure 3 *Burrows of pholads in wood*
(Photo courtesy of Dr Simon Cragg)

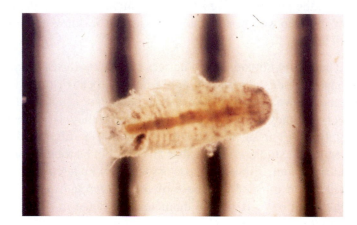

Figure 4 *Light micrograph of Limnoria sp.*
(Photo courtesy of Dr Simon Cragg)

note are *Paralimnoria* and *Phycolimnoria*. The adult animals are whitish grey in appearance and are quite small, ranging in length from 2–4 mm (Figure 4). The galleries and tunnels they excavate at or just below the wood surface are 1–3 mm in diameter and the resulting network of tunnels has a lacework appearance. Sloughing off of the surface remnants produces an eroded appearance to badly damaged wood which is commonly most pronounced in the intertidal zone.

Limnoriids are highly mobile. The animal has a head, a segmented thorax and abdomen, with seven pairs of legs positioned ventrally under the thorax. Pleopods positioned under the abdomen function as gills and are used in swimming. The last two posterior segments of the animal – the fifth pleomer

Figure 5 *Adult Sphaeroma triste*
(Photo courtesy of Dr Simon Cragg)

and the pleotelson – have ornamentations on the dorsal surface that are used for identification.

The limnoriids have a world-wide distribution, but water temperature has a major influence on the occurrence of particular species, their reproduction and their rate of attack. The warm-water species *Lygropia tripunctata* is most active at 22–26°C and reproduction occurs at temperatures above 14°C. The adult females carry the fertilised eggs and developing juveniles in a ventral brood pouch below the thorax before they are eventually released into the excavated galleries. The cold-water species *Limnoria lignorum* is only active below 20°C and begins breeding at 9–10°C. At temperatures below 10°C, the rate of wood degradation declines significantly. Salinity is also a factor that affects limnoriid activity and animals are able to tolerate only 40–50% normal seawater concentrations for short periods of time.

Sphaeromatids. These are often referred to as pill bugs because of their capacity to curl up into a ball when they are disturbed. Like the limnoriids, they are isopods and have a comparable segmented body comprising a head, thorax and abdomen. However, they are significantly larger than limnoriids with mature sphaeromatids reaching 10–15 mm in length (Figure 5). Such relatively large wood-boring animals are capable of causing rapid damage to timber structures in seawater by burrowing across the grain, perpendicular to the wood surface in the inter-tidal zone.

The burrows they produce are *ca.* 5 mm in diameter and 15 mm or more in length, so each animal is thought to excavate its own burrow to more or less fill the space it has created. Although it is thought that animals leave their burrows to reproduce, it has been suggested that pairs can live together maintaining their broods at the end of a shared burrow.

The ability of sphaeromatids to digest wood fragments excavated during burrowing is unclear, and although gut cellulases have been found, the amount of wood removed is thought insufficient to sustain the development of one animal. Food is more likely derived from planktonic filter feeding and grazing on microorganisms on the surface of the burrows. In this respect, the sphaeromatids have a lifestyle comparable to the pholads, *i.e.* using the burrow as a safe retreat rather than as a source of food.

Active wood-destroying sphaeromatids are found in warm temperate–tropical waters. They exhibit tolerance to a range of environmental extremes, including short periods of desiccation in hot conditions when wood is exposed, plus the ability to withstand very low salinity conditions for days at a time.

Chelurids.   The crustacean family Cheluridae is perhaps the least studied of the marine wood-borers. These amphipods are pale yellow to red in colour, with a size range of 3–8 mm in length. In some respects, their morphology is not unlike that of the limnoriids, with a head and segmented thorax, but the posterior abdomen of the animal is comprised of fused segments and has a different form and appearance (Figure 6). Chelurids have been described as the scavengers that live off the waste produced by the limnoriids. They are invariably associated with limnoriid infestations feeding on faecal pellets produced by them as well as the wood surfaces of the limnoriid galleries. Like the limnoriids, the chelurids copulate in the excavated galleries and the female carries the eggs and developing larvae in a ventral brood pouch. However, when the young chelurids are released, they search out established galleries

Figure 6 *Light micrograph of Chelura sp.*

rather than excavate their own tunnels. The close association between these two groups of wood-borers not only offers the chelurids a suitable habitat in which to live, but is also a factor in promoting sexual maturity in the chelurid population.

Chelurids are widely distributed, and *Chelura terebrans* is found in wood in temperate and subtropical waters. Wood infested by both limnoriids and chelurids normally has the chelurid population closer to the surface while the limnoriids inhabit the deeper region. Chelurids are also less tolerant of environmental extremes and do not survive long in conditions of low salinity or low oxygen.

2.2 Lignolytic Marine Microorganisms

As mentioned earlier, the major agents of damage to archaeological timbers in oxygenated marine environments are the marine wood-boring animals that destroy the wood over relatively short periods of time. In contrast, microbial decay of wood in these situations is slow and progressive and in anoxic environments such as sediments, it may take centuries for degradation of the polysaccharide components of the wood cell walls to occur in large timber structures. As with attack by animals, the speed of microbial degradation will depend on the natural durability of different wood species, the greater resistance of heartwood versus sapwood, plus environmental factors including water temperature, oxygen availability and salinity. A further consideration in this regard is the role of microorganisms in the digestion of wood fragments in the gut of marine-boring animals and the symbiotic relationship that exists between some members of these two groups of organisms. This will be dealt with in more detail later, but suffice to say that the accepted view that wood biodeterioration in the sea is a marine-borer problem may have underestimated the importance of microorganisms in the overall process.

For many years, bacteria have been recognised as early colonisers of damp and wet wood. In terrestrial situations, they are regarded as the first phase of a consortium of decayed microorganisms most of which have greater lignolytic capabilities and include the soft rot fungi (Ascomycota and mitosporic fungi) and the brown and white rot fungi (Basidiomycota). In the marine environment, submerged wood is colonised quite rapidly by bacteria and soft rot fungi, but colonisation by basidiomycetes is less common.

Bacteria and soft rot fungi are more tolerant of the low oxygen conditions in saturated wood, and in wood that is buried in sediments bacterial decay tends to predominate. Early reports identified unicellular bacteria in foundation piling and shipwreck timbers, but more recently three bacterial decay types in wood cell walls are now recognised – erosion, tunnelling and cavitation bacterial decay.

2.3 Erosion Bacteria

Erosion bacteria are considered the dominant microbial degraders of archaeo-logical, waterlogged wood. It is now accepted that waterlogged wood main-tained under conditions of minimal oxygen availability is decayed primarily by erosion bacteria although some attack by soft rot fungi and tunnelling bac-teria can be found in these timbers. It has been suggested that the very low oxygen conditions are periodically alleviated for this to occur.

Decay by erosion bacteria has been reported in different wood species result-ing in the utilisation of wood polysaccharides; the extent of lignin modification is not known. This type of degradation is characterised by the formation of ero-sion troughs or grooves on the lumen surface of wood cell walls (Figure 7) and is most effectively visualised using scanning electron microscopy (SEM). The eroded areas are associated with single-celled bacteria that are typically rod shaped and decay is progressive from the lumen surface towards the compound middle lamella of the cell wall. Initially, a population of erosion bacteria can be observed on the lumen surface, sometimes orientated in a parallel fashion fol-lowing the alignment of the cellulose microfibrils. Grooves excavated in the S_3 layer of the wall by individual bacterial cells are more or less the same size as the cells, indicative of restricted enzyme activity around each bacterium.

Attachment to the lumen surface is through the formation of a glycocalyx and as progressive breakdown occurs, the secondary cell wall is converted

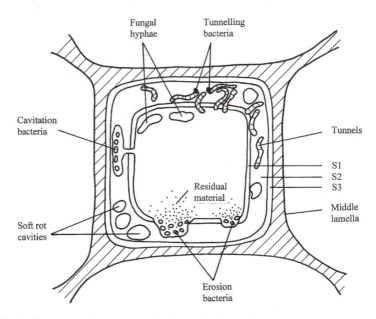

Figure 7 *An illustration showing microbial attack of wood*
(After Singh and Kim, 1997)

into amorphous material mixed with bacterial cells and bacterial slime. It can therefore be concluded that extracellular mucilage is important in bacterial cell adhesion, motility and enzymatic degradation of the cell wall. In softwood timbers, attack of the cell wall by erosion bacteria can also result in the formation of angular cavities in the S_2 layer and this is probably caused by interaction between the orientation of cellulose microfibrils and extracellular bacterial enzymes. So far, identification of erosion bacteria from field specimens has not been successfully accomplished. A greater understanding of their physiological requirements in laboratory culture and the optimum conditions for promoting adhesion and decay at cell wall surfaces is necessary for this to be achieved.

2.4 Tunnelling Bacteria

Decay of wood cell walls by tunnelling bacteria is found in wood exposed to the sea. This type of decay was also recorded in saturated wood exposed in the warm water of industrial cooling towers. In both situations, oxygen is limiting, but does not approach the near-anaerobic conditions of burial in silt and sediments. The presence of tunnelling bacteria in wood in soil contact is indicative of the ubiquitous nature of these organisms and their ability to decay preservative-treated wood and durable heartwood is further evidence of the aggressive capabilities of these wood-decay organisms.

Tunnelling bacteria remove the polysaccharide components of wood cell walls and some of the lignin. The appearance of the decayed wood is soft and darkened. Longitudinal sections of the decayed wood viewed with a transmission light microscope, display regions of granulation in softwood tracheids. Fine tunnels are found at the margin of these regions and with appropriate staining and good microscope optics, single-celled bacteria can be seen at the ends of the tunnel. The granulation zones develop from a loose, irregular network of fine tunnels following bacterial penetration into the wood cell wall (Figure 8). Under polarised light, the zones of granulation show loss of birefringence confirming the breakdown of the crystalline cellulose in these infected areas. But not all examples of tunnelling bacterial decay show exactly the same pattern of attack. In some cases the tunnels appear wider, are fewer in number and appear to radiate more precisely from a central point producing an 'ice-fern' pattern.

Decay patterns associated with tunnelling bacteria are not uniform and differences in their appearance may be influenced by environmental factors, wood species or indeed by the bacteria themselves, since to date none have been identified.

The penetration of these bacteria into wood cell walls is initiated by adhesion to the lumen surface via an extracellular glycocalyx. The bacterium lyses the S_3 layer and is then able to tunnel in the S_2 and S_1 layers. The direction of tunnelling

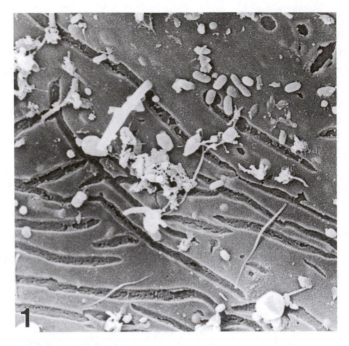

Figure 8 *An SEM of bacterial attack (erosion) of wood*

is apparently random and transmission electron microscopy (TEM) shows bacteria penetrating through the primary wall and middle lamella into adjacent cell walls. In badly attacked regions, the cell wall is riddled with tunnels, showing bacteria at their distal ends and a series of concave transverse layers along the tunnels. These transverse layers or cross walls are thought to be composed of mucopolysaccharide deposited by the advancing bacterium and have been described as chambered tunnels. These observations suggest that cell wall lysis is therefore not a continuous process.

Recent studies into the condition of waterlogged archaeological wood from a range of sources, indicate that attack by tunnelling bacteria and soft rot fungi is more concentrated in the outer layers and erosion bacteria predominate deeper into the wood, suggesting the latter are much less oxygen dependent.

2.5 Cavitation Bacteria

Bacterial attack of wood cell walls, which produces cavities can take two forms, tunnelling – one chambered (see above) and the other non-chambered tunnelling or cavitation attack. Cavitation attack was described from preservative-treated pine posts in horticultural soils and microscopically, the decay pattern was similar to that described in ancient softwood piling timber in Stockholm. Decay occurs in the wood cell wall producing angular, often diamond-shaped cavities. Attack is initiated at the lumen surface where bacteria attach themselves via extracellular slime. Following penetration of the S_3 layer, the underlying S_2 is attacked resulting in the formation of small cavities. It is assumed that one bacterial cell or a small group enlarges the cavity and as cavities increase in size they begin to coalesce. Although the bacterial population in the cavities will increase, it is assumed that enzyme breakdown of the cell wall is non-localised and that cells produce diffusible substances similar to those observed with brown rot decay. It is clear from the literature that any definitive diagnosis of cavitation attack by bacterial cells can only be achieved with the aid of SEM and TEM facilities.

2.6 Bacterial Symbionts of Shipworm

Apart from recent observations of bacterial endosymbionts associated with the pholad *Xylophaga* (see above), the most detailed studies of symbiotic associations in marine-borers have focussed on the teredinids. Electron microscopy pinpointed the gland of Deshayes, a specialised region of the shipworm gill, as the site where bacterial symbionts reside. Unlike many symbiotic associations, the bacteria can be isolated from the animal and cultivated in the laboratory and so far symbionts from seven shipworm host species have been maintained in culture. The bacteria produce cellulolytic and proteolytic enzymes, plus nitrogenase for atmospheric nitrogen fixation allowing the shipworm to survive principally on a wood diet that is naturally nitrogen depleted. Investigations into the identities of bacterial symbionts from four shipworm species representing three different genera, revealed a phylogenetically common bacterial strain *Teredinibacter turnerae*, a member of the gamma subdivision of the *Proteobacteria*. However, later studies on the identity of the bacterial symbiont from *B. setacea*, found that this teredinid harboured a different bacterial symbiont. This work also showed that acquisition of the symbiont by successive generations of *B. setacea* was through a vertical mode of transmission, the symbiont cells being present in the host ovary and offspring.

As a footnote to this work, the gut contents of *Limnoria* have been examined using SEM but no microorganisms were associated with wood particles even though the wood ingested by the limnoriids was decayed by soft rot fungi and tunnelling bacteria. This suggests that ingested microbes were digested by the animals.

2.7 Soft Rot Fungi

Although soft rot decay of wood was first described in the 1950s, prior to that date numerous authors had recorded similar symptoms in decayed wood from a range of aquatic and terrestrial habitats. The decay is superficial and below the surface the wood is invariably sound. Soft-rotted wood is characteristically darkened in colour and has a spongy soft texture when wet. A further characteristic is the extensive longitudinal and cross-cracking that occurs at the surface when the wood dries out. Wood exposed in seawater suffers from attack by soft rot fungi. The speed of infestation and decay of the wood depends on its natural durability, whether the material is sapwood or heartwood, and on the environmental conditions at the site. Following submergence, the conditions within the wood, *i.e.* water saturation and reduced oxygen levels, promote the growth of lower ascomycete and mitosporic fungi. Members of these two groups of fungi are traditionally recognised as agents of soft rot decay, but the development of microscopic features characteristic of soft rot have also been reported in wood infected by basidiomycete fungi.

It is now estimated that there are 444 formally described higher marine fungi and of these, 360 species from 177 genera belong to the Ascomycota, 74 species from 51 genera belong to the mitosporic fungi and 10 species from seven genera belong to the Basidiomycota. A great majority of these fungi appear to have adapted to an aquatic habitat by virtue of the elaborate appendages on their spores. The spore appendages are not only considered important in aiding buoyancy and therefore dispersal in the sea, but they also have an important role in aiding attachment of spores to suitable substrata, notably wood surfaces. Following attachment and spore germination, colonisation and decay by these fungi will take place. Studies to assess the decay capabilities of marine fungi suggest that around three-quarters of these could cause mass loss in wood.

The soft rot decay process begins with hyphal colonisation of wood via the rays and in hardwoods also via the vessels. Soft rot can take two forms: cavity attack (type 1) and erosion of the cell wall (type 2). Cavity attack is the classical form of soft rot and when viewed in longitudinal sections of decayed wood, appears as chains of biconical or diamond-shaped cavities, and more or less cylindrical cavities with conical ends (Figure 9). Entry into the wood cell wall by hyphae growing in the lumen of the wood cell occurs through transverse penetration of a fine hypha into the S_3 layer. On reaching the S_2 layer, the fine penetration of hypha is realigned parallel to the direction of the cellulose microfibrils. The fine hypha grows in an axial direction, but eventually apical growth ceases. Enzymes released from the hyphal surface start to degrade the wood cell wall creating a cavity, and gradually the width of the hypha increases. After a short period of time, a second phase of apical growth begins with the emergence of a fine hypha from the tip of the cavity. This hypha then grows in the cell wall for a period of time, but again growth ceases. Cavities develop in the cellulose-rich S_2 layer of the secondary wall and follow the helical orientation of the cellulose microfibrils; in badly attacked wood, a series of cavity chains aligned in parallel can be observed before the cell wall begins to collapse. Chains of cavities are therefore formed as a result

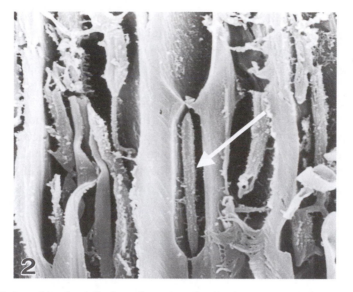

Figure 9 *Soft rot fungal attack of wood in cell wall. Note fungal hyphae in cavity (arrowed)*

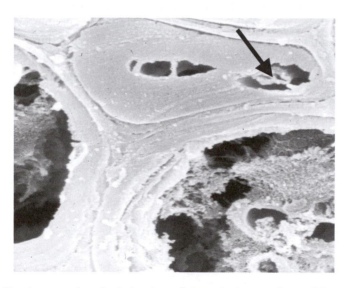

Figure 10 *Transverse section of oak showing soft rot cavity in secondary wall layer (arrowed)*

of 'start–stop' apical growth of fine hyphae combined with cavity expansion due to cell wall dissolution by enzymes released from the hyphal surface.

When viewed in transverse section, the cell walls of axial cells are perforated with holes in the S_2 layer that represents cavities in cross-section (Figure 10). Hyphae within the cavities continue to release enzymes and the cavities widen, coalesce and eventually break through the S_3 layer into the cell lumen. In severely decayed wood, the high-lignin-containing compound middle lamella is the only remnant of the cell wall.

Although soft rot cavities are commonly visualised in the axial tracheids of softwoods and the fibres of hardwoods, the walls of other cells such as vessels and parenchyma also show cavity attack.

Erosion of wood cell walls by soft rot fungi is less commonly reported. Hyphae growing in the cell lumen release enzymes that produce erosion patterns on the cell wall surface. These can take several forms including V-shaped nicks or smooth erosion troughs around the hypha, but in general, this type of attack is more prevalent in susceptible hardwoods than softwoods.

2.8   Basidiomycete Fungi

Decay of wood by lignicolous basidiomycete fungi in terrestrial situations is generally described as either brown rot or white rot, although occasionally there are reports of basidiomycetes producing soft rot cavities. The terms brown rot and white rot reflect the appearance of badly decayed wood. This distinguishes between those fungi that destroy polysaccharides rapidly, while lignin depletion is slow – brown rot fungi, and those that are able to decompose all of the structural wood components (cellulose, hemicelluloses and lignin) – white rot fungi. The brown rot type of decay has so far not been recorded for marine fungi.

Of the 10 species of marine basidiomycetes recorded, three have been tested for their decaying ability. These are *Digitatispora marina*, *Halocyphina villosa* and *Nia vibrissa* – all white rot fungi. Using three wood species – balsa (*Ochroma lagopus*), Scots pine (*Pinus sylvestris*) and beech (*Fagus sylvatica*), wood blocks were inoculated with the three basidiomycetes as well as the soft rot fungus *Monodictys pelagica*. This work is of particular interest because three of the fungi – *D. marina, N. vibrissa* and *M. pelagica* were isolated from the 450-year-old timbers of the Tudor ship *Mary Rose*. While balsa is a very low density, highly perishable timber species and the high weight losses were not too surprising, it is interesting that significant weight loss values were recorded in beech after 24 weeks incubation. Although beech is a perishable timber, the ability of these fungi to cause significant weight loss in so short a time, in archaeological terms, is a good indicator of their long-term decay ability. For the record, oak is listed as durable in the Durability Classification.

White rot attack occurs at the lumen surface of the wood cell wall where hyphae adhere to the wall. Enzymes released along the hyphae create discrete erosion troughs around the hyphae with mucilaginous material extending into the troughs; a further feature is the presence of bore-holes resulting from hyphal penetration through the cell walls. As with the soft rot fungi, decay by basidiomycetes is dependent on the availability of sufficient levels of oxygen to sustain growth and decay ability. Studies with terrestrial white rot fungi have shown that lignin degradation is a highly oxidative process and it can be assumed that this is true of basidiomycete decay of wood in the sea.

2.9   Insect Borers

One important biodeteriogen that can cause damage to recovered archaeological timbers is the insect *Nacerdes melanura* (Oedemeridae) – the wharf-borer. This insect has a world-wide distribution and is found in freshwater inland sites and damp timbers in terrestrial sites as well as marine pilings, wharves, jetties, *etc.* The larvae that are 12–30 mm in length, burrow into both hardwood and softwood timber, but have a preference for softwoods (Figure 11). Although the wharf-borer is generally regarded as a pest of damp decayed timber, larvae do infest wood at or just above the high-water mark and have been observed burrowing into sound wood in close proximity to decayed wood. During the conservation programme for the Tudor ship *Mary Rose*, larvae were found burrowing in stored waterlogged archaeological timbers. Larvae were found in oak, poplar and pine timbers, but those affected made up only 2% of the total examined. The moisture contents of the timbers ranged between 131 and 670% and larval burrows were found to a depth of *ca.* 1 cm, where soft rot and bacterial decay were present.

Another potential pest of timbers in maritime situations is the wood-boring weevil *Pselactus spadix* that can excavate galleries in the inter-tidal and splash zones of wharf timbers. The larvae are believed to be the primary tunnellers,

Figure 11 *Wharf-borer adult beetle emerging from a hole in the wood surface*

but since adults are found in the galleries as well, it is not clear how the weevils survive inundation of the wood, albeit briefly, and how they deal with the high salt content in the wood consumed.

3   WOOD STRUCTURE

Throughout history, wood has been a natural, renewable resource with a range of properties and uses. The characteristics of timber are defined by its cell components and the arrangement of these components, forming a three-dimensional structure that has strength, flexibility and decorative appeal. The first important distinction to be made when considering the use or value of timber is the source; *i.e.* is the material a hardwood, derived from deciduous, broad-leaved trees, or is it a softwood, derived from evergreen, coniferous species. Although this rule is not totally universal, for the most part the distinction holds true and it allows us to recognise anatomical features and cell types that are quite characteristic of hardwoods or softwoods. A second consideration when examining timber for particular purposes is the distinction between sapwood and heartwood. These features are often easily seen on the exposed face of a crosscut log where commonly the lighter coloured, outer sapwood band can be distinguished from the often darker inner heartwood. The significance of these two zones is the difference in durability, since the heartwood has greater natural resistance to damage by marine animals, insects and microorganisms, by virtue of the significantly higher levels of chemical extractives in the tissue. It is also important to recognise that different wood species have different levels of natural durability based on the performance of heartwood exposed to ground contact. The time taken to fail in soil contact provides a class index of durability, ranging from very durable to perishable species.

The terms hardwood and softwood do not necessarily reflect the strength, engineering or durability properties of a particular wood species, for instance yew (*Taxus baccata*) that is a softwood species, is extremely dense, strong and highly durable. Wood density is therefore an important consideration when assessing the strength and long-term performance of a species and in the case of hardwoods is imparted via densely packed axial bundles of fibres. Fibres are elongate, thick-walled, cylindrical cells with pointed ends having a very narrow central lumen running the length of the cell. Their vertical orientation within the trunk of the tree, alongside wide, open, thin-walled vessels that have a water-conducting function, and axial parenchyma and fibre-tracheid cells, make up the axial elements of the trunk and the converted timber. In addition to the axial system, there is also a radial system of cell elements that is composed of ray parenchyma cells, forming uni- or multi-seriate sheets of cells orientated vertically and extending from the outer cambial zone below the bark, to the centre of the trunk. It is significant to mention that in addition to the vertical and radial orientation of cell types there is the possibility for water movement through narrow pits in the wood cell walls.

In softwoods, the major difference to the anatomy of the wood is the absence of vessels and fibres that are replaced by elongate axial tracheids. Tracheids conduct water and provide strength to the timber and are characterised by having elaborate bordered pits in the cell wall. Tracheid cell wall thickness changes through the growing season so that in the early part, the wall is comparatively thin with a wide lumen whereas later in the season the walls are much thicker and the lumina much narrower. This can be seen with the naked eye as representing annual growth rings in cross-cut logs. Like hardwoods, softwoods also have a radial system of cell elements composed of ray parenchyma and ray tracheids and the bordered pits provide a similar avenue for water movement in the tangential direction. In effect, wood might be described as a labyrinthine network of microscopic tubes and galleries connected together via fine holes in the walls.

Wood cell walls have an outer primary and an inner secondary wall layer. The thin primary wall is formed first and is flexible to accommodate changes to the shape of the cell as it grows and develops. The secondary wall is thicker and more complex and is laid down when the cell shape is established. In addition, the secondary wall is composed of three layers – the outermost S_1, adjacent to the primary wall, the S_2 and the innermost S_3, which are adjacent to the cell lumen. Cells are held together by the middle lamella that cements the outer surface of primary wall layers of one cell to the primary walls of adjacent cells.

The three major chemical components of wood cell walls are cellulose, hemicelluloses and lignin, but of these only cellulose performs a structural function by providing strength to the material. Cellulose is a long-chain homopolymer composed of thousands of glucose monomers bonded together by regular $\beta1$–4 chemical linkages. The molecular organisation of cellulose in wood cell walls is based on aggregations of cellulose chains into microfibrils that adopt a helical orientation within the cell wall. The angle of the helix is different in different cell wall layers, but in the S_2 layer the angle is particularly steep. In addition, the S_2 layer is thicker than the other layers and is extremely cellulose rich. Unlike cellulose, hemicelluloses are short-chained, branched heteropolymers composed of several different monosaccharides as well as glucose units. The branched nature of the molecules that are about 200 units in size, requires three different types of bonding and hemicelluloses are considered to be part of the matrix in which cellulose microfibrils are embedded. However, the enzymatic breakdown of hemicelluloses bears close similarity to that of cellulose degradation through the activity of hydrolase and oxidase enzymes.

In contrast to the two-dimensional nature of the wood polysaccharides, lignin is described as a three-dimensional polymer composed of phenyl propane units. It is a recalcitrant molecule that is degraded slowly in nature and along with the hemicelluloses, is a major component of the matrix material in wood cell walls. Lignin is a complex molecule and in addition to the phenyl propane monomers has up to 15 different types of chemical bond to hold the polymer together. Its distribution within the wood cell wall is not uniform and lignin is found in greatest concentration in the middle lamella region and the cell corners. Hence, the progressive enzymic destruction of wood cell walls leaves these more highly lignified regions as the final remnants of the structure.

Clearly the physical and chemical composition of woody materials has a direct bearing on the ecology of the organisms that destroy them. The lignin matrix is, in some measure, a barrier to the activities of microbial cellulolytic and hemi-cellulolytic enzymes except in the case of basidiomycete white rot attack where the three wood cell wall components are broken down. However, in the thick cellulose-rich S_2, where lignin is more or less absent, cellulase enzymes are able to function freely. All wood-destroying micro-organisms are capable of degrading cellulose, but for some their status as lignin degraders is unclear. This is especially so for lignolytic bacteria, although using TEM, tunnelling bacteria have been visualised traversing lignin-rich middle lamellae.

3.1 Archaeological Wood Recovered from the Marine Environment

The marine environment presents a hostile and seemingly unlikely situation for the survival of archaeological wood, yet it does survive. Normally, wood does not survive long enough in marine environments to enter the archaeological record because of the activities of wood-boring animals and aerobic microbes. However, studies have shown that rapid burial in the anoxic sediments of the seabed will protect ships' timbers and wooden artefacts from the physical, chemical and biological processes that influence the deterioration of exposed wood.

Waterlogged wood recovered from the marine environment is usually fragile due to the degradation of the cell wall components and is supported mainly by seawater. Some heavily degraded waterlogged archaeological wood may be unable to support its own weight in air. In a wet condition, preserved archaeological wood can retain its dimension and shape. However, despite the well-preserved appearance at excavation, recently excavated archaeological waterlogged wood does not dry as sound wood. When dried, the wood cells weakened by degradation are often unable to resist drying stresses and collapse, and surface cracking and splitting may occur resulting in the destruction of an entire historic artefact.

3.2 Properties and Condition of Waterlogged Archaeological Wood

The result of microbial decay in the marine environment is a progressive structural weakening of the wood due to the sequential loss of important polymers. Chemical analysis of waterlogged archaeological wood shows a significant loss of hemicellulose and cellulose, whereas the lignin component remains more or less unaffected. Small wooden artefacts appear to deteriorate almost uniformly through their entire thickness. Larger artefacts are frequently found to have a heavily degraded surface layer, surrounding a core, which is sound. These important differences in decay pattern are reflected in a classification scheme for waterlogged archaeological wood. This scheme gives a simple description of a wood sample in terms of loss of solid component. The most heavily degraded wood is considered to be class I (Figure 12), and the least degraded is class III, as shown in Figure 13. Surprisingly, it is the class III timbers where some of the biggest conservation problems occur.

Chemical hydrolysis can also cause loss of hemicellulose and cellulose, but this type of degradation is not usually significant in comparison to biological attack. These processes have been implicated in the loss of carbohydrates from wood preserved in cold anoxic environments. The fastest chemical change to archaeological wood is that of solvent swelling that is caused by the cellulose molecules becoming surrounded by a sheath of hydrogen-bonded seawater. This results in increases in size of about 12% tangentially, 5% radially and 2% longitudinally. Chemical changes also occur as a result of mineral

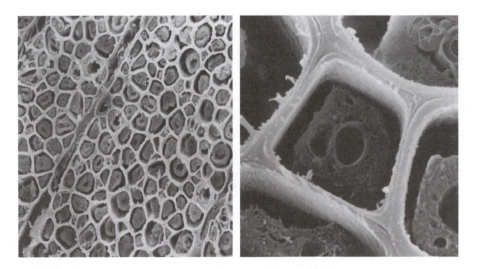

Figure 12 *An SEM of class I condition of waterlogged archaeological oak. Note the heavily degraded cell wall layers*

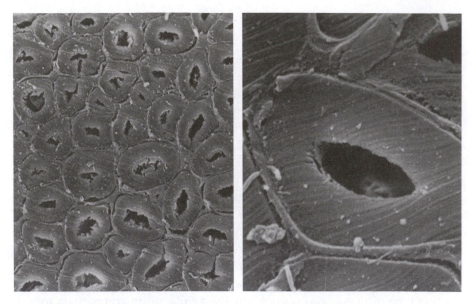

Figure 13 *An SEM of class III condition of waterlogged archaeological oak. Note the*
well-preserved cell wall layers

inclusion. Sea salt is passively introduced in marine environments, iron(III) is actively chelated by cellulose and tannates, and high concentrations of iron corrosion products build up as a consequence. As the wood becomes anoxic, sulfate-reducing bacteria (SRB) convert these salts to metal sulfides. A physical consequence of mineral inclusion is that the pores between the lumina are blocked, and the wood becomes impermeable.

3.3   Storage of Waterlogged Archaeological Wood

In comparison to aerobic environments, deterioration of archaeological wood in the anoxic sediments of the seabed occurs at a slow rate. Buried waterlogged archaeological wood will suffer little deterioration, for as long as the anaerobic environment remains undisturbed. When removed from its burial environment, physical, chemical and biological degradation ensues and the wooden artefact must be recovered and stored or conserved soon after exposure. The waterlogged state of archaeological wood must be maintained on its removal from the marine environment until it is conserved.

The fundamental problem encountered if waterlogged archaeological wood is allowed to dry in an uncontrolled manner, is that it may shrink, collapse, distort, split and even in severe cases, completely disintegrate. The situation becomes a particular problem if the recently recovered wood is stored under conditions that do not prevent further deterioration. In an ideal world, it is probably worth aiming for a storage period of less than 12 months. This is not always possible when dealing with a well-preserved marine wreck site that can provide many tons of well-preserved waterlogged wood. In such cases, storage periods of up to 20 years are not uncommon.

During storage, waterlogged archaeological wood is subject to attack by bacteria, fungi and insects. To prevent such biodeterioration in waterlogged archaeological wood, a storage method should be employed which must fulfil the following criteria: (i) maintain maximum moisture content; (ii) prevent or minimise further decay; (iii) be compatible with subsequent studies and treatment method; (iv) be easy to maintain; (v) be inexpensive; and (vi) must be non-toxic for conservators who maintain it and subsequently work on it. Of greatest importance is that the wood will always be kept water saturated from the time of recovery until conservation treatment begins. Numerous methods have been employed to store waterlogged archaeological wood:

(i) wrapping in polyethylene bags,
(ii) immersion in tanks containing water,
(iii) spraying with water, and
(iv) keeping wood buried in anoxic sediments.

3.4 Wrapping in Polyethylene Bags or Sheeting

The most commonly used method of storage is wrapping wooden artefacts or individual ships' timbers in polyethylene bags of various sizes. Polyethylene sheeting is often used and all edges are hermetically sealed to prevent the drying of the stored artefact. Wrapped wooden artefacts are often kept within buildings where there is no temperature control. Maintaining the waterlogged state of the wooden artefact or timber is an important criterion and should not be underestimated. If wood is maintained in a waterlogged condition, the growth of wood-rotting fungi may be restricted since water saturation will limit the amount of available oxygen that is required by the fungi. However, low oxygen tensions may not totally prevent the growth of certain fungi, especially soft rot fungi. These fungi are better adapted to decay wood in conditions of low oxygen tensions, since they preferentially degrade cellulose through a non-oxygen process. Examination of polyethylene-wrapped *Mary Rose* timbers has been demonstrated to support wood decay microorganisms. Fruiting structures of 17 fungi were found on the surface of timbers. Three of these were terrestrial fungi and 14 were marine fungi (see Table 1). The most frequent species was *M. pelagica* recorded on 23% of timbers examined.

Despite the large number of wood-decay fungi growing on the surface of stored timbers, few signs of recent attack were observed. Soft rot cavities were observed but were not numerous. Such decay patterns were only observed

Table 1 *Marine fungal species isolated from*
Mary Rose timbers

Species	Decay type
Ascomycotina	
Corollospora maritima	Soft rot
Lulworthia sp.	Soft rot
Neriospora cristata	Soft rot
Zopfiella sp.	Unknown
Deuteromycotina	
Cirrenalia macrocephala	Soft rot
Dictyosporium toruloides	Unknown
Gliomastix mucorum	Unknown
Humicola alopallonella	Soft rot
M. pelagica	Soft rot
Penicillium sp.*	Staining
*Stachybotrys atra**	Staining
*Trichocladium achrasporum**	Soft rot
Zalerion maritima	Soft rot
Zalerion varium	Soft rot
Basidiomycotina	
D. maritima	White rot
N. vibrissa	White rot

* Terrestrial species.

after extensive examination of a large random timber sample, indicating that fungal attack of stored polyethylene-wrapped timbers was very localised.

Bacterial decay was more frequently observed. Once again this form of decay was also very localised. Erosion and tunnelling decay patterns were observed in a number of timbers. The erosion troughs were observed below the outer decayed surface layers, while the tunnelling bacteria were observed closer to the outer cell layers. The possible explanation for the lack of active decay in polyethylene-wrapped stored wood may be simply that unfavourable conditions exist in timbers that are encapsulated in polyethylene. Oxygen levels within the bags are very low, with little gaseous exchange across the polyethylene barrier. Also, with little or no exchange between the polyethylene barrier system and the external environment, a build up of toxic substances and secondary metabolites takes place, thereby inhibiting the growth of fungi and aerobic bacteria. The combined action of low oxygen and the presence of these substances may explain the lack of recent decay by aerobic microorganisms in polyethylene-wrapped timbers.

Breaches in the wrapping material may allow colonisation by wood-boring insects. A survey of 1,568 *Mary Rose* timbers stored in polyethylene bags revealed the presence of wharf-borer beetle larvae in 2% of these timbers. Attack by this insect is not always apparent from examination of the timber surface. The presence of flight holes or small piles of frass (insect droppings) was often the only visible indication that the timbers were infested. Damage caused by the wharf-borer larvae is often extensive. They produce tunnels throughout the softer decayed regions of the timbers. Study of polyethylene-wrapped timbers also indicated that the adult beetle could penetrate the poly-ethylene enclosure. To prevent infestation of insects, stored archaeological wooden artefacts should be wrapped in a metallised polyester barrier film and all edges hermetically sealed.

Biocides have been employed in attempts to control insect and microbial activity in polyethylene-wrapped timbers. However, their limited efficacy over time makes re-treatment necessary. This is often time consuming and creates future problems for conservators in the handling of the wood and in the safe disposal of the biocide solution. It must also be noted that biocides may inter-fere with radiocarbon dating and future active conservation treatment processes. Alternatively cold storage of polyethylene-wrapped wood revealed that the activity of wood-degrading microbes is arrested. However, the viability of the bacteria and fungi was not affected by cold storage. The threat of renewed decay under more favourable conditions therefore exists. Cold storage and biocide treatments are therefore only suitable for the short-term storage (less than 1 year) of wrapped timbers where the risk of biodeterioration is con-sidered low. A number of *Mary Rose* wrapped timbers were stored in a CO_2 atmosphere. This method was ineffective in preventing wood decay activity, or the viability of the mixed microbial flora supported by wrapped archaeo-logical timbers. During a study to evaluate the efficacy of this method, the anaerobic environment created was found to support the growth of anaerobic human pathogens, presenting a serious health risk to conservators. In addition, many fungi are able to grow at very low oxygen tensions, thus presenting a biodeterioration risk to the wood if complete anaerobiosis is not achieved and maintained in storage.

In many archaeological projects involving marine wrecks of historic import-ance, the temporary storage of waterlogged wood all too frequently turns into longer-term storage while conservation issues, such as funding, are debated or deferred. Extending passive storage indefinitely may lead to further decay of the wooden artefact over a period of several years until it is in too poor a condition for active conservation. To prevent this situation from developing, conservators have used gamma irradiation to control bacteria, fungi and insects living in poly-ethylene-wrapped waterlogged wood. It is suggested that gamma irradiation of waterlogged archaeological wood at a dose sufficient to kill biodeteriogens but not have adverse effects on wood quality could be an effective passive conser-vation treatment for the long-term storage of waterlogged wood. The application of irradiation as a means of killing microorganisms and insects has been exten-sively researched in the field of conservation. A number of workers have successfully used irradiation to sterilise archival and manuscript papers. Decreases in mechanical strength of cellulose-based papers were observed after gamma irradiation at a dose of 10 KGy. Nevertheless, the advantage of irradiating material after packaging has been recognised in the food industry.

Gamma rays are emitted from the radionuclides of cobalt 60 and caesium 137. Caesium is a fission product obtained from spent nuclear fuel rods. This must be extracted in processing plants before being used as a useful source of radiation. Cobalt 60 is manufactured specifically for commercial use in radiotherapy, radiography and food processing. It can be produced in nuclear reactors in large quantities by neutron bombardment of inactive cobalt 59 for periods usually in excess of 6 months. Almost all gamma ray facilities in the world use cobalt 60 rather than caesium 137. The half-life of cobalt 60 is 5.3 years. The source must be changed periodically to maintain a given level of radioactivity. Ionising radiation is only safe for use at energies below 5 MeV, since radiation at higher energies may make materials such as wood radioactive. Gamma radiation, however, does not result in any residual radioactivity in foods since cobalt 60 has an energy of only 1.17/1.33 MeV.

Research at the *Mary Rose* Trust has shown that gamma irradiation is a superior alternative to low temperature and biocide treatments of polyethylene-wrapped timbers and has been adopted by the Trust since 1998. Screening a range of bacteria and fungi isolated from waterlogged archaeological wood revealed that a dose of 15 KGy is required for the inactivation of most organisms (see Table 2).

Gamma irradiation may also be desirable as a treatment for wood excavated from polluted archaeological sites. Applying medical sterilisation standards of 25 KGy ensures complete inactivation of any human pathogens in addition to wood decay organisms. A dose of 15–25 KGy had no adverse effects on the chemical and physical properties of waterlogged archaeological wood. However, at dose levels in excess of 100 KGy radiolytic damage was observed resulting in increased hygroscopicity, warping, loss of surface texture, reductions in compressive and bending strength, and chemical alteration and degradation of wood cell wall components.

Table 2 *Radiosensitivity of wood decay organisms*
            (After Pointing, 1995)

Organism	Data source	Recommended lethal dose (KGy)
Aerobic bacteria	Mary Rose timbers	2.4
Anaerobic bacteria	National Collection of Industrial Bacteria	2.5
Invertebrates (insects)	Mary Rose timbers	3.2
Marine fungi	Mary Rose timbers	12.0
Terrestrial fungi	Mary Rose timbers	13.0

Advice to Conservators. Gamma irradiation, is suitable for the passive treatment of both slightly- and heavily-degraded waterlogged wood. An absorbed dose of 100 KGy is acceptable for any one timber, and this dose level should not be exceeded in a single or repeated treatment, since higher irradiation dose may adversely affect the physical and chemical nature of the wood. One drawback to this process is that gamma-irradiated timbers are unsuitable for radiocarbon dating and any analytical procedure is invalidated by an exposure to irradiation.

Gamma irradiation treatment involves wrapping the waterlogged wooden artefacts in a metallised polyester barrier film and transporting them to a commercial irradiation facility. For *Mary Rose* timbers, inactivation of wood biodeteriogens was achieved by a single dose of 20 KGy. Following irradiation treatment, it is important to maintain the integrity of the wrapping material surrounding the sterile wooden artefact. Any breach will allow contamination of the sterile environment. One of the advantages of gamma irradiation is that there is no residual radioactivity in the treated artefact, which is completely safe to handle. This presents a clear advantage over biocide treated objects that often present toxicity problems. Gamma irradiation is also a low-cost alternative to cold storage, with far greater efficacy in preventing any on-going decay. Finally, gamma irradiation may be compared favourably to the criteria for an ideal passive conservation treatment method (see Table 3).

3.5 Storage of Large Wooden Artefacts

Wooden artefacts too large to enclose in a wrapping material have to be passively held without suffering serious drying effects. Tanking large items of artefacts, such as boats and individual ships' timbers, is a practical way of storing large waterlogged archaeological wood. The size of individual artefacts will dictate the size of water storage tanks used. Plastic tanks with lids to exclude

Table 3 *An evaluation of gamma irradiation as an ideal passive holding method*

Criteria for ideal storage	Evaluation of gamma irradiation
Prevent or minimise further decay	Complete sterility of artefacts
Compatibility with future treatment	At recommended dose levels (below 100 KGy), no adverse effects
Ease of maintenance	No specialised environment needed but wrapping material must remain intact
Cost effectiveness	Wrapping and irradiation cost, low storage and maintenance costs
Health and safety	Do not result in any residual activity in the exposed wood

light are often used for the storage of smaller items. The advantage of this type of container is that it can be placed in a cold storage room. For larger wooden artefacts (gun carriages, *etc.*) recovered from marine wreck sites, purposely built tanks are required and are constructed from a variety of material, such as polyethylene, GRP and stainless-steel tanks. These tanks are lidded to prevent light penetration. All water storage tanks are usually kept inside a building to prevent exposure to temperature extremes.

Although this method is successful in preventing moisture loss from stored waterlogged archaeological wood, growth of slime and wood degrading organisms is often observed. From time to time it is necessary to control the growth of microorganisms within storage tanks. Cool temperatures alone are not able to prevent slime formation, but can help in retarding its development. Bio-control using snails (*Physa* sp.) or fish (perch) has proved effective in controlling bacterial slimes, while ultraviolet light performs relatively poorly. The poor penetrative powers of ultraviolet radiation limit its application to surface sterilisation of water-stored timbers, but ultraviolet radiation has been used as a means for reducing algal growth in freshwater storage tanks containing heavily degraded archaeological wood.

Biocides have also been employed in attempts to control microbial growth within water storage tanks. However, the hazardous nature of biocides to the conservator and environment and their limited efficacy over time makes re-treatment necessary at regular intervals. When considering the use of biocides, several factors need to be considered: (i) effectiveness, (ii) resistance to biocide, (iii) toxicity and (iv) compatibility with the stored artefactal material. Often biocides tend to be more effective on certain groups of organisms. Before choosing a biocide, the agents of decay should be identified and a suitable concentration to control the problem determined. Some organisms such as bacteria have the ability to become resistant to a biocide (usually a genetic ability). This problem of resistant strains of bacteria emerging often necessitates a change in the type of biocide. A biocide should not pose a health hazard to operatives applying it and should be compatible with future conservation treatment methods. A list of biocides used in the control of microbes and insects include phenols, chlorophenols, orthophenyl phenol, thymol, amphoteric surfactants (Tego), salicylanilides, quaternary ammonium compounds, hypochlorites and boric acid/borax.

More recently, metal ions and ion complexes (copper:silver ion disinfection technique) have been used to disinfect conservation water storage tanks containing waterlogged archaeological wood. The special properties of copper as a bacteriocide, fungicide and an algicide, and silver as a bacteriocide are well documented. Copper and silver are introduced into the tank by the ionisation of Cu:Ag alloy electrodes and it is recommended that the treatment solution does not exceed 7.4 pH or there is the likelihood of the copper coming out of solution. Finally, water storage of archaeological wood should not exceed a period of 3–5 years.

3.6   The Storage of the *Mary Rose* Hull

With large structures such as the *Mary Rose* hull, where dismantling is impractical, tank storage methods can be difficult to achieve due to the size of the tank needed and the inaccessibility of the wooden structure once in storage. In such instances, direct spraying of the wooden surface is employed. The spraying of ships and other large structures with freshwater has an added advantage in that the sediments and salt deposits are washed out. However, a major problem with the passive storage of a large structure such as the *Mary Rose* hull is one of size. Conservators have developed storage methods for small finds and most are not capable of expansion to a larger scale. Thus, the only suitable method to passively store the hull was spraying. A research programme was set up to compare a warm (20°C) and a cold (5°C) spraying regime using original ships' timbers. The data obtained were predictable, showing significant differences in bacterial activity between the higher and lower temperatures. Maintaining a water spray temperature of between 2 and 5°C substantially reduced the degree of microbial activity within hull timbers. Based upon these findings, a low-temperature spraying regime was adopted by the *Mary Rose* Trust. The reasons were as follows: (i) water saturation of the hull timbers would be maintained by the constant flow of water; (ii) temperature of the spray water would be maintained between 2 and 5°C (at these temperatures, microbial activity is well below the optimum growth temperature of marine and terrestrial fungi and wood-degrading bacteria); (iii) the method is safe; and non-toxic and (iv) this method is compatible with future conservation treatment methods.

Following its recovery in October, 1982 (Figure 14), the entire hull was sprayed with chilled fresh water. Using this method the hull was passively held for a period of 12 years during which post-excavation recording, cleaning and reconstruction work were completed. Throughout the period of passive storage, the water temperature was maintained between 2 and 5°C. During this period, the activity of microbes was evident and from time to time it was necessary to treat the timber surfaces with a biocide due to the build up of slime (bacteria, fungi, diatoms and green algae). Although bacteria and fungi were found to grow superficially on hull timbers, little active decay of the ships' timbers was observed. Microbes isolated from hull timbers during the passive storage period included species of the bacteria *Aeromonas, Alcaligenes, Pseudomonas* and *Vibrio* and fungal species identified included *M. pelagica, Lulworthia* sp., *H. alopallonella, N. cristata, C. macrocephala* and *Tricladium* sp. The optimum growth for these bacterial and fungal species isolated from hull

Figure 14 *Spraying of the hull of the Mary Rose with chilled fresh water*

timbers was found to be between 20 and 30°C. This supported the view that suppression of microbial activity could be temperature controlled.

Throughout the passive holding period, there was a real risk of bacterial infection to staff working within the hull environment. Many common aquatic bacteria are able to colonise man-made water spray systems. Poorly maintained water systems provide ideal sites for their growth and *Legionella pneumophila*, a common aquatic bacteria, was considered a possible hazard. The strategy adopted to control this bacterium and others was maintenance of low water temperature, maintaining a clean spray system, carefully monitoring bacterial populations and the selective use of biocides to reduce the growth of slimes. Adhering to these simple guidelines removed the risk of infection to both public and staff.

3.7 Reburial

The feasibility of reburial as a storage method for waterlogged archaeological wood has been fairly well used. The near neutral pH, low Eh and absence of dissolved oxygen fulfil the requirements assumed by conservators for the ideal burial environment to minimise decay. Unfortunately, when shipwrecks are reburied there is little or no subsequent monitoring to determine effectiveness of such a technique. Before contemplating reburial as an alternative storage technique, substantial work is required to establish whether wood is susceptible to anaerobic decay. However, any decay by the anaerobic cellulolytic flora in marine sediment is likely to occur slowly. A further consideration when contemplating reburial is the speed at which the disturbed marine sediment returns to fully anoxic conditions, since re-exposure of wood to aerobic seawater/sediment will result in the onset of decay by bacteria, fungi and marine wood-borers. It is also necessary to monitor the burial environment throughout the period of storage. Periodic inspection of buried timbers or buried test wood samples, and monitoring of the sediment microcommunity is also necessary. There are few international reburial projects being conducted at the moment, which include on-site environmental monitoring and safeguarding. These studies are all concentrated in the northern hemisphere in marine environments.

The long-term survival of wood in marine sediments is dependent on the conditions prevailing in the burial environment. Considerations include the Eh of the sediment and pore water, the type and stability of the sediments, the chemical composition of the pore water and seawater and the level of biological activity in both seawater and sediment. Careful monitoring of the changes in these important environmental parameters over time will ensure the success or failure of reburial as a long-term storage method. Primarily reburial leads to an initial increase in oxygen levels and water content allowing biodeterioration to continue for a short period until anoxic conditions are re-established. This is due to the microbial processes within the sediments, which consume the oxygen leading to the suppression of aerobic microbes.

The depth of burial is also important when considering reburial of archaeological wood. Studies have shown that shallow burial will not prevent long-term deterioration of buried wood by microbes but at depths greater than 50 cm the extent of deterioration decreases significantly.

If reburial in anoxic marine sediments does prove to be successful as a passive storage method of waterlogged archaeological wood, it is likely to be applicable only to long-term storage due to the expense and efforts in carrying out reburial and site monitoring.

The natural degradation of wooden shipwrecks *in-situ* can be slowed down, or even eliminated, by covering the site with a physical barrier such as polyethylene or polypropylene textile (Figure 15). This method is very effective because it uses the wrecked environment to create an overlying mound of sediment, which becomes part of the site. It also physically prevents colonisation of any subsequently exposed wreck timbers by wood-boring animals, like shipworm.

Figure 15 *Covering of exposed archaeological timbers with Terram 4000 (geotextile). Note the growth of marine plants on Terram surface* (After Pournou, 1999)

4   CONSERVATION OF ARCHAEOLOGICAL WOOD

The appearance of waterlogged archaeological wood can be highly deceptive, often disguising the fact that part or the entire artefact may now be a highly-deteriorated matrix of lignin, carbohydrate debris and minerals, which is supported by water. The chemical changes that occurred during exposure to the underwater marine environment will have invariably weakened the wooden object, making it extremely susceptible to ambient conditions. Without active conservation, irreversible shrinkage will take place and destroy the object.

There are two chemical problems that need to be addressed during the treatment of waterlogged archaeological wood. First, the degradation of chemical components of a wooden object is the direct result of long exposure in both seawater and anoxic marine sediments. Second, the object will have become contaminated by minerals (mainly iron and sulfur compounds). The first problem is considered irreversible, involving the structural breakdown of hemicelluloses and cellulose (Figure 16) by chemical and biological processes. Conservation treatments of ancient waterlogged wood, therefore aims to control these processes as the object is transferred from the water-logged state to a dry and stable condition.

Conservation in the context of waterlogged archaeological wood must remove mineral inclusions and stabilise the size and shape of the object. In addition, the conservation treatment should be reversible, and give long-term stability. All treatments of waterlogged archaeological wood should adhere

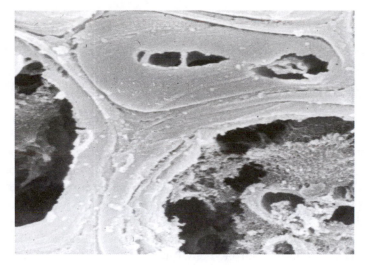

Figure 16 *An SEM of oak fibre showing decay of the secondary wall layers (hemicellulose and cellulose component). Note the residual lignin component in the cell*

to the principle of minimum intervention, and the best conservation method is one that interferes with or modifies the wood to the least extent.

4.1 Definitions and Aims

The types of chemical treatments involved in the conservation of archaeological wood are: (i) lumen-filling treatments that fill the spaces within the wood with an inert chemical to provide structural support and prevent collapse, (ii) bulking treatments that enter the cell walls and reduce cell wall shrinkage, and (iii) surface coatings that cover the surface of a dry object.

4.2 Lumen-filling Treatments

Large amounts of a high molecular weight chemical are introduced into the cell lumina and permanent voids (Figure 12) to improve the physical properties of the object and to prevent cellular collapse of very decayed wood cells. This treatment must form a solid in the cell lumina and voids as water is removed.

However, there is one major drawback to lumen-filling treatments in that they can cause osmotic collapse if too great a concentration is employed across the wood cellular structure. Since the mass transfer of a chemical into wood is proportional to this concentration gradient, the result is an extremely slow process. Calculating a maximum sustainable concentration gradient, and length of time taken for an object to equilibrate at each concentration step, is one of the most fundamental issues that conservators face when devising a lumen-filling treatment method. Previously, this often involved a certain amount of guesswork, but more recently conservators have experimented by nuclear magnetic imaging to help solve these problems.

Lumen-filling chemicals used by conservators include high molecular weight, solvent-based polymers such as styrene, methyl methacrylate and butyl methacrylate. These are usually polymerised by heat or gamma irradiation. The problem with these chemicals is that they produce a composite object in which a significant part is the polymer. This often results in a considerable increase in object density.

A more suitable lumen-filling treatment is high molecular weight polyethylene glycols (PEG) (1500–4000). This polymer is often used to treat highly degraded waterlogged archaeological wood. A major drawback with the use of high grade PEG is that the object tends to look waxy and wet, due to a reduction in the intensity of internally-reflected light from cells near the surface of the object. Furthermore, unlike the solvent-based polymers, PEG-treated objects must be kept in controlled environments (18–20°C, 55% RH).

4.3   Bulking Treatments

Bulking treatments are used to reduce cell wall shrinkage of well-preserved wood cells (Figure 13). Only small molecules are able to penetrate intact cells and are not suitable for the stabilisation of highly degraded archaeological wood. These treatments also act indirectly by reducing the vapour pressure of the water present in the microcapillary system of the cell walls, and so retain water in these areas. This helps to keep the cells turgid, and so provides hydraulic support as the cell dries.

Bulking chemicals can be divided into non-reactive and reactive. Non-reactive chemicals enter into the molecular structure of the cell walls, replacing water, and they help maintain cell size and shape. These chemicals react with remaining cellulose by means of hydrogen bonding. Examples of non-reactive chemicals include low molecular weight PEGs (200–600) and sugars. Non-reactive chemicals also reduce the stiffness of the cell wall layers. In theory, these can be removed from the wood matrix by warm water extraction.

Reactive bulking chemicals form covalent bonding in the cell wall. Examples include alkylene oxides and low molecular weight thermosetting resins. Unlike non-reactive chemicals, treatment with these chemicals is considered non-reversible.

4.4   Surface Coatings

Small amounts of chemicals are often applied to the surface of dry archaeological wood. Coatings are used to improve the surface quality and integrity of very fragile objects. Waxes (beeswax) and PEG 6000 are frequently used to coat dried wooden objects.

5   CONSERVATION METHODS

The following discussion deals with different conservation methods under a number of headings in relation to their application for the conservation of small and large wooden objects. The main headings comprise: (i) PEG treatments, (ii) *in-situ* polymerisation, (iii) freeze drying, and (iv) air drying. Before commencing a treatment it is necessary to remove or neutralise contaminants that enter the object through long-term exposure to seawater and the burial sediments.

5.1   Removal of Mineral Inclusions

Salts are passively absorbed into wooden objects in seawater, but some ions notably iron(III) are actively bound by cellulose, and high concentrations of iron corrosion products build up in consequence. As the wood becomes anoxic, these salts are converted into sulfides by the activity of SRB. A physical consequence of iron inclusions is that the wood microstructure becomes blocked making the object impermeable to future conservation treatments. Iron sulfide is unstable in the aerobic environment, and its oxidation can destroy the object (Figure 17).

50mm

81A3882

Figure 17 *Kidney dagger handle from the Mary Rose with salt formation on surface*

As a consequence, the presence of high concentrations of iron sulfides in dry conserved archaeological wood results in structural damage and blooms of sulfate crystals develop on the object surface. In addition, the sulfuric acid produced inside ship timbers and artefacts starts to hydrolyse the wood. An analysis of *Mary Rose* timbers has shown high and variable iron concentrations, up to about 8% in the surface layers. Also, iron ions catalyse oxidative degradation of cellulose in a chain reaction, involving free radicals and molecular oxygen, and the wood loses much of its strength.

To ascertain the composition of iron/sulfur-containing particles in archaeological wood, the pre-edge features of iron and sulfur K-edge X-ray absorption near-edge spectroscopy (XANES) and X-ray diffraction has provided valuable information on the iron compounds and its oxidative state. Iron compounds commonly encountered in archaeological wood are goethite (FeOOH), magnetite (Fe_3O_4), molysite ($FeCl_3$), pyrrhotite (FeS), pyrite (FeS_2), mackinawite (Fe_9S_8), rozenite ($Fe^{II}SO_4 \cdot 4H_2O$), melanterite ($Fe^{II}SO_4 \cdot 7H_2O$), and natrojarosite ($Fe^{III}_3(SO_4)_2(OH)_6$).

It is imperative that iron compounds are removed from waterlogged archaeological wood. In order to dissolve iron compounds from archaeological wood, special complexing agents (chelates) are used to form strong and also soluble complexes with iron(III) ions. Ethylenediiminobis(2–hydroxy–4–methyl–phenyl)acetic acid (EDMA) is a chelating agent used by scientists at the *Mary Rose* Trust.

In conjunction with XANES and X-ray diffraction, initial investigations involving synchrotron based X-ray microspectrometry (at ESRF) has revealed two types of reduced sulfur compounds in the timbers of the *Mary Rose*. Organosulfur compounds are found in the lignin-rich middle lamella between wood cell walls, mostly as thiols, disulfides and elemental sulfur, while inorganic iron sulfides including iron pyrite, occurs in separate particles in the cell lumen.

Based upon these finding, new methods are now being developed to control the acid-forming oxidation processes, by removing reactive iron sulfides and stabilising the organo-sulfur compounds. Current ideas indicate that after removing the reactive iron sulfides, the remaining organo-sulfur compounds could be prevented from forming acid by the addition of antioxidants to conservation treatments, enabling long-term stable conservation of marine archaeological wooden objects.

By using sulfur spectroscopy, synchrotron-based XANES and X-ray microspectrometry it has been revealed that the acid problem originated from microbial activity within the marine sediments (Figure 18).

SRBs are known inhabitants of marine sediments. These bacteria convert sulfates into sulfides under anaerobic conditions. Sulfides in the form of hydrogen sulfide penetrate the wood and react to form two categories of sulfur

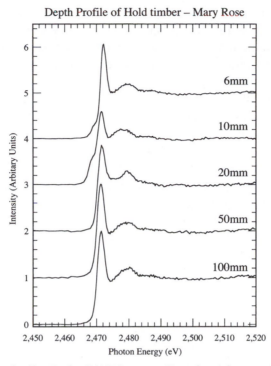

Figure 18 *Normalised sulfur K-edge XANES spectra (Daresbury) from sections of oak core from surface (0 mm) and inwards taken from untreated Mary Rose timber*

compounds, which can be treated differently: particles of iron(II) sulfide present in the cell lumen and organosulfur compounds firmly bound in the lignin-rich parts (middle lamella) of the wood cell wall. The sulfides are oxidised in humid conditions and are probably the source of most of the acid present. In the timbers of the *Mary Rose*, sulfides are abundant, but there is little sulfate, probably because of the constant spraying during the active conservation treatment.

Monitoring the state of archaeological wood and its environment during conservation is an important aspect for their long-term preservation. Changes of environment during excavation, treatment and storage can result in the initiation of chemical and biological processes that may ultimately destroy the object. The acidification of archaeological wood once excavated is a problem of immediate concern. It is therefore important to monitor the chemical changes and microbial activity in the timbers during conservation and display.

5.2 Microbial Activity

Sulfur-metabolising organisms have been monitored within the timbers of the *Mary Rose* by extraction and analysis of bacterial ribosomal RNA. The diversity of bacteria identified comprised sequences representing bacterial groups capable of sulfur oxidation (*Thiobacillus prosperus*, *Methylobacterium* spp. and pseudomonads). They are capable of transforming reduced sulfur compounds present in archaeological wood to sulfuric acid.

A possible treatment under investigation for sulfur contamination in *Mary Rose* timbers involves the controlled oxidation of reduced sulfur species by microbes. Timber samples previously examined by XANES have been treated with a chemical bio-stimulant containing a powerful blend of six specially selected non-pathogenic strains of *Bacillus*, a non pathogenic chemolithotroph, *Thiobacillus denitrificans* and two Gram-negative sulfur-oxidising facultative heterotrophs of the species *Starkeya novella* and *Paracoccus pantotrophus*. The end product of their activity is sodium sulfate instead of sulfuric acid, which can be easily washed out from archaeological timbers.

In addition to this study, timber samples previously examined by XANES and infused with acid salts have been treated with various biocides to inhibit continued bacterial oxidation of reduced sulfur species. This may also help prevent continuing production of acid.

To prevent oxidation of reduced organosulfur in the wood cell walls, conservators are currently investigating antioxidants. Commercial antioxidants (ascorbic acid, anthocyanidins, proanthocyanins, flavon-3-ols, flavonones and flavonols) are being used to prevent oxidation of reduced species by mopping up free radicals.

The removal of sulfur contamination from wood is not easy. To date conservators have tried various solvents for removing sulfur compounds, including alcohol, acetone, and tertrachloroethylene (TCE). Hydrochloric acid has been used to remove acid-volatile sulfur (AVS), essentially pyrrhotite. However, after extensive extraction of hydrochloric acid-treated samples with the best elemental sulfur solvent, TCE, S k-edge XANES of the samples found little change in the spectrum, although there may have been some change in the total sulfur content. Chromium(II) reducing agents have also been used to remove elemental sulfur and pyrite, and we are currently working on ways to measure the chromium-reducible sulfur (CRS) and AVS in samples of archaeological wood; these methods are currently in use in soil science. However, these methods of removing sulfur compounds are not suitable for the treatment of actual artefacts. It seems unlikely that any extraction process will be possible, at least on any large scale.

However, it is worth noting that while the Swedish warship *Vasa* is still contaminated with sulfur and generating acid, it is now apparent that during its treatment by spraying a considerable amount of acid was generated, as the pH of the spray solution kept falling. This was neutralised by the addition of borax/boric acid buffer. In the case of the *Vasa*, the major sulfur contaminant appears to be elemental sulfur that is essentially insoluble in water. However, borax is a well-known detergent, as well as a biocide, and scientists have found that it increases the solubility of elemental sulfur in water. It may be that the elemental sulfur, once solubilised in water, becomes more readily oxidisable. Unfortunately, in the case of the *Vasa*, this process was not carried to a suitable endpoint, principally because the problem was not then recognised; it was assumed that the acidity was caused by breakdown of the PEG (formation of formic acid).

5.3 Polyethylene Glycol Method

By far the most commonly-used method for conserving small and large water-logged wooden objects is the application of PEG. PEGs are polymers of ethylene oxide with the generalised formula of $HOCH_2(CH_2OCH_2)_nCH_2OH$. All PEGs are designated with a number that represents their average molecular weight and these ranges from 200 to 20,000. Low molecular weight PEGs (200–600) are clear viscous liquids at room temperature, and are readily miscible with water. Higher molecular weight PEGs (>1000) are white waxy solids at room temperature, with decreasing solubility in water and solvents as molecular weights increase.

PEGs are considered inert and safe. They are used in a great variety of applications because of their chemical structure, low toxicity, solubility in water and their lubricating properties. Since PEGs are considered to be so versatile, conservators have for the past 4 decades recommended their use in the treatment of waterlogged archaeological wood. The advantages of using PEGs outweigh their disadvantages.

It is well known that PEG-treatment methods require long treatment times and increasing hygroscopicity with decreasing molecular weight can be a problem when treated objects are stored under high relative humidities ($>60\%$). This has led to the frequent use of high molecular weight PEGs (1500 and 4000). Unfortunately, these high molecular weight PEGs require very long treatment times before sufficient levels can be detected inside the wood structure.

Although there are variations in the use of the PEG method between conservation laboratories, conservators usually follow the same basic procedure. A wooden object is placed into a tank or container containing an aqueous solution of low concentration PEG at an elevated temperature of 30–60°C. The concentration is gradually increased by the evaporation of water or by the addition of more PEG, until 70–100% PEG concentrations are reached for highly degraded wood or around 15–25% concentrations for sound archaeological wood (followed by freeze drying). Excess PEG is removed and the PEG-treated timbers are vacuum freeze dried or air dried. In some laboratories, the process has been accelerated by dissolving PEG in solvents or methanol, again with some level of success for relatively small objects. When considering the use of PEG, four methods of application can be considered: (i) tank method for complete structures (Bremen Cogg – Bremen, Germany), (ii) tank method for individual units or components, (iii) spraying of large structures (*Vasa* – Stockholm, Sweden and Mary Rose – Portsmouth, UK), and (iv) surface treatment (dry objects only).

PEG: Two Stage Method (Case Study – The Mary Rose Hull).   Research by Trust scientists explored in great detail the physical and chemical characteristics of the hull timbers and quickly discovered that they were differentially decayed. The outer surface to a depth of 10 mm was soft and heavily degraded, while wood cells beneath this region were found to be well preserved. According to shrinkage data, there can be as much as 25% shrinkage to the wood structure as the timbers dries out. Drying the hull without some form of stabilisation was considered unsuitable and the use of PEG was investigated.

The wide variety of properties of PEGs has complicated conservation efforts in the past. For example, low molecular weight PEGs could move swiftly and efficiently into the wood cell walls, helping to stabilise them and prevent cell wall shrinkage. But they are hygroscopic and give a tacky feeling. Thus, until recently, conservators would only use molecular weight of 1500–4000; they might not have a rapid uptake of PEG but they would not be left with a tacky surface.

To overcome this problem, conservators and scientists began to experiment with a two-stage treatment process involving low and high molecular PEGs. This research programme indicated that a lower weight PEG (mw 200) to start with, followed by a higher molecular weight (mw 2000) was the best combination for the differentially decayed timbers of the *Mary Rose*. The treatment programme began in 1994 (Figure 19) with the entire hull surface

Figure 19 *Spraying the hull timbers with a solution of PEG*

being sprayed with an aqueous solution of PEG 200). The starting concentration was 10% and over a 10-year period was raised to a final concentration of 40%. In 2006, the hull timbers will be sprayed with a 40% solution of PEG 2000 to consolidate the very decayed outer timber surfaces. This will take approximately 5 years to achieve optimum levels for surface stabilisation. In 2011, the PEG treated timbers of the *Mary Rose* hull will be air dried.

5.4 *In-situ* Polymerisation with Radiation Curing Monomers and Resins

As an alternative to PEG treatment of waterlogged wood, attempts have been made to use a radiation curing method involving a range of monomers and resins. The aim of such treatments is to reduce treatment times and achieve a more stable artefact. Styrene, vinyl acetate, acrylonitrile, acrylates and methacrylates are the most studied monomers in the treatment of waterlogged archaeological wood. The most widely used of these monomers is the water-soluble vinylpyrrolidone and methacrylamide and the non-water soluble chemicals such as n-butyl methacrylate monomer or unsaturated polyester resin.

For water-soluble chemicals, there is direct water monomer exchange, while the non-water soluble chemical is dehydrated by means of a solvent such as acetone and then immersed in a solution of the monomer. After a suitable impregnation period, polymerisation is induced by exposure to gamma rays emitted by a cobalt 60 source. To date the results achieved for either water-soluble or solvent-soluble monomers have not given satisfactory results. Methods using monomers and gamma irradiation require highly-specialised applications and the process is very costly. In general, such methods should only be adopted if PEG treatment cannot be used or if waterlogged wooden objects are heavily contaminated with iron and/or sulfur compounds.

5.5 Drying Following Conservation Treatment

The removal of water from saturated wooden objects without destroying the object is a significant challenge for conservators.

Air Drying. This is the simplest drying technique in which water undergoes a phase change to the vapour state, and is removed as such. This technique can result in unacceptable amounts of damage to a waterlogged wooden artefact because of excessive cell wall shrinkage and collapse. Some type of treatment is usually applied beforehand to prevent the dimensional response of an object to drying. Both bulking and lumen-filling treatments are used before air drying an object. An object having a highly deteriorated surface with a well-preserved core is usually pre-treated with a combination of a bulking agent (PEG 200–600, 30–40% w v^{-1}) and a lumen-filling treatment (PEG 2000–4000, 70–100% w v^{-1}).

Recent investigations at the *Mary Rose* Trust have shown that no advantage would be gained from adopting a complex schedule for the drying of PEG-treated archaeological wood. Use of high initial relative humidities is likely to result in slightly elevated dimensional change and may cause unwanted migration of low molecular weight PEG from the core to the outer surface. Drying temperatures above 20°C will also cause small increases in wood shrinkage and collapse. Current data suggests that the most suitable drying schedule would be initiated at 55% relative humidity and this level uniformly distributed around the drying object. Drying directly to 55% RH at 18°C is also likely to have additional and financial benefits.

Recent studies have also highlighted the importance of tightly controlling environmental conditions after drying. Humidity levels higher than the recommended 55% RH are thought to have contributed to the formation of sulfuric acid within the timbers of the Swedish warship, the *Vasa*. The air-conditioning system in the display hall has now been upgraded to cope with this problem.

Vacuum Freeze Drying.   One way of eliminating surface tension forces during drying is to freeze the object, and then remove the frozen water within the wood by sublimation. This is the most commonly-used drying technique for treatment of pre-treated waterlogged archaeological wood. However, this technique has the disadvantage of requiring the water within the object to undergo a two-phase change as shown in Figure 20.

As water freezes it expands by approximately 9%, so a cryoprotectant such as PEG 200–600 is required to protect the object during freezing. As small amounts of low molecular weight PEGs are required, the process of freeze drying is very much faster than the consolidation process required by air drying. Within a frozen PEG-treated object, a complex mixture of phases can exist. As the temperature is lowered, there is a tendency for the material with the highest freezing point (water) to freeze first, possibly with some cryoprotectant as a eutectic. In theory, a whole range of different compositions could freeze out, one after another; more usually most material will freeze as a glass. As water is subsequently sublimated out from the wooden object, the cellulose matrix of the wood cell walls will come to favour PEG. The ideal of freeze drying is not, therefore, attained, as the drying front will be in the liquid phase. When all of the pure ice has sublimated, the vapour pressure of the sample will fall, and there is some risk that the remaining eutectic will warm up above its melting point and this would be detrimental to the sample. Despite these theoretical problems, freeze drying is more reliable and much faster than air drying. The freeze drying facility at the *Mary Rose* Trust has a capacity of 3 m^3 and has been used with excellent results.

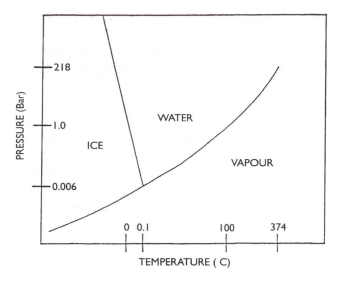

Figure 20 *Phase diagram of water*

6 CONCLUSIONS

With the passive storage of waterlogged wooden artefacts, the major problem is not that we do not have adequate storage methods available, it is in recognising the limitations of these methods with respect to the size of the artefacts that can be treated within the restrictions of time, funding and facilities available. Such considerations, in conjunction with data available from numerous experimental investigations should be taken into account when deciding upon the most suitable conservation method for the short- and long-term storage of waterlogged archaeological wood.

Of the methods discussed, gamma irradiation is the most effective method for inactivating wood-decay organisms, and is the only method that provides complete sterilisation. The surface dose for most timbers of the size usually stored by wrapping is 20–30 KGy and it is the only treatment that meets the criteria for an ideal storage method. For large intact structures such as wooden ships and boats, a low temperature spray is the most successful storage method tested. Methods other than the use of biocidal chemicals to maintain clean water for spraying, include the use of fish (species of perch) and snails in water storage tanks.

Reburial of waterlogged wood in the marine environment is now becoming increasingly common practice in managing wooden historic wrecks *in situ*, predominantly due to the prohibitive conservation costs. Remedial strategies that are appropriate and cost-effective are fundamental pre-requisites for the successful long-term management of underwater wreck sites.

With the active conservation of waterlogged archaeological wood, the major problem is not that we do not have adequate treatment methods available, it is in recognising the serious limitations of these methods. However, those methods involving PEG offer the greatest flexibility and variability.

Dehydration techniques and methods involving *in situ* polymerisation are not considered suitable for anything other than small wooden artefacts. Freeze drying offers the conservator a safe method of drying small PEG-treated objects while air drying is more suited to the drying of larger structures, such as intact ships if the condition allows it.

Several problems remain to be solved before effective and new conservation methods can be developed. New treatments are being developed aimed at controlling acid-forming oxidation processes by removing the reactive iron sulfides and stabilising the organosulfur compounds. In this regard, the removal of sulfides and iron compounds from the timbers of the Swedish warship *Vasa* poses an important and immediate challenge to conservators charged with the responsibility of maintaining this significant monument in a stable condition for posterity.

REFERENCES AND FURTHER READING

M. Jones (ed), *For Future Generations: Conservation of a Tudor Maritime Collection*, Cromwell Press, Trowbridge, England, 2003.

C. Pearson (ed), *Conservation of Marine Archaeological Objects*, Butterworths, London, 1987.

R.M. Rowell and R.J. Barbour (eds), *Archaeological Wood, Properties, Chemistry and Preservation*, American Chemical Society, Washington, 1990.

M. Sandstrom, Y. Fors and I. Persson, *The Vasa's New Battle: Sulfur, Acid and Iron*, The Swedish National Maritime Museums, Stockholm, Sweden, 2003.

CHAPTER 12

In-situ Preservation of Waterlogged Archaeological Sites

DAVID GREGORY AND HENNING MATTHIESEN

Department of Conservation, Archaeology, National Museum of Denmark, PO Box 260 Brede, DK-2800 Kongens Lyngby, Denmark

1 INTRODUCTION

Increasingly, newly discovered archaeological sites are being preserved *in situ*, as opposed to being excavated. Many factors are influencing this trend, including the high costs of recovery coupled with the conservation and subsequent curation and management of recovered remains in perpetuity. Furthermore, it may not always be desirable, or necessary, to excavate a site. Yet, as part of the cultural resource it should be preserved or managed *in situ*. Thus, there is an ever-increasing need to develop an alternative to conventional conservation. *In situ* preservation is a form of preventive conservation and has become a common solution to the management of archaeological sites in wetlands and on underwater sites. This solution is, however, only viable if we are able to understand the threats against the archaeological remains and control the environment at the site. Deterioration cannot be completely avoided and therefore *absolute* preservation *in situ* is not achievable. All sites are dynamic, continuing to form in the sense that deterioration processes keep on altering the material remains *albeit* at slow, often imperceptible, rates. However, conditions conducive to the preservation of archaeological artefacts can be assured by a sequence of environmental monitoring combined with the analysis of artefacts from the site and the study of the deterioration of analogous modern materials. In this way the processes of deterioration, state of preservation and the future threats to artefacts, or a site, can be identified. This will enable a risk assessment of a site to be made. Based on this, there are three options: Passive preservation, *i.e.* leave the site as it is, the environment is safe; Active preservation, *i.e.* influence environmental parameters around the site in order to enhance preservation conditions; Excavation, *i.e.* mitigation strategies will be too expensive or ineffective and the best option to ensure that archaeological information survives would be to excavate and conserve or re-deposit (re-bury) artefacts in an environment more conducive to their preservation. It is important to remember that *in situ* preservation is not a *panacea* and may not be feasible in all situations. However, approaching and assessing a site in this way can provide a more objective rationale for the management of archaeological sites.

This chapter provides a recommended approach to *in situ* preservation of waterlogged archaeological sites containing wooden and iron artefacts, and complements the chapters on deterioration of wood and metals. It draws on research carried out at the waterlogged site of Nydam, an Iron Age sacrificial site in southern Denmark, which has been preserved *in situ* since 1997. The techniques used will be outlined, and an overview of the results and their implications will be given. Although Nydam is a waterlogged site, the recommended framework is applicable to underwater sites as the processes of deterioration are the same even though the specific agents of deterioration may differ.

2   ENVIRONMENTAL MONITORING

Waterlogged soils in wetlands are normally characterised by reduced conditions and the lack of oxygen from a certain depth. Apart from this common feature they cover a large range of pH values and concentrations of nutrients and other species. These parameters may influence the deterioration of archaeological artefacts, and it is thus necessary to monitor the conditions at each individual site in order to evaluate the feasibility of *in situ* preservation. A range of different methods have been used for environmental monitoring at archaeological sites in waterlogged soils.

2.1   Water Level

The hydrology of a site is very important, in particular the water level around artefacts. If, for example, wooden artefacts lie above the water table they will be exposed to deterioration by a whole range of micro- and macro-organisms, which will lead to their complete destruction within years. More importantly, depending upon their state of preservation they may suffer from collapse with a resultant loss of archaeological information. Water level is normally measured using dipwells/piezometers. Dipwells are the simplest form of piezometer and consists of an open-ended tube, which is either perforated near the base of the tube or along the length, and is inserted into a borehole. Measurements of water level in a dipwell can be made using a "dipmeter", consisting of a probe on a graduated cable or tape measure, which gives an audible signal when the probe comes into contact with water. Another method is simply to use a piece of plastic tubing taped to the side of an extending ruler and lower this into the dipwell. By blowing through the tubing, bubbles can be heard when the end of the tube is in contact with the water in the dipwell. Other methods, which have successfully been used are datalogging devices, which give continual measurement of the water level. Dataloggers are expensive but give a better understanding of the dynamics of the system, for instance during particularly wet or dry periods.

2.2 Pore Water Composition

Literature on the biodeterioration of organic materials and the corrosion of metals show that parameters such as pH, alkalinity, conductivity, oxygen, sulfate, sulfide, ammonium, nitrate, nitrite, phosphate, potassium, chloride, iron and manganese may be relevant to monitor. Some of the parameters are suspected to affect the deterioration of archaeological materials directly, while others reflect the ongoing chemical and biological processes in the soil. Following the creation of a waterlogged environment, oxidised species within the soil rapidly decrease due to their utilisation by chemical reactions and microorganisms. The lack of dissolved oxygen suppresses the activity of aerobic microorganisms. Thus, there is an increase in the activity of anaerobic microorganisms, which in turn leads to the production of chemical species in their reduced form. For example dissolved oxygen will tend to be depleted within the first few centimetres of soil below the water level. Nitrate, sulfate and ferric iron levels decrease, and an increase is seen in their reduced forms such as ammonium, sulfide and ferrous iron.

Ideally oxygen, along with other parameters, should be measured directly *in situ* in the soil. The electrodes used for oxygen measurements directly in the soil have to have extremely low or no oxygen consumption, otherwise they may use up oxygen faster than it is supplied in the soil, thus always giving a zero oxygen reading after a short while. Clark-type microelectrodes, which were developed for measurements in sediment, have been used with success, due to their fine size (sub-millimetre) and design they do not require stirring to obtain reliable readings (Figure 1). With this concept, it is possible to get an immediate idea of the oxygen concentrations in soil. It can give a very detailed picture of the vertical variation at a site with a minimum of physical disturbance. The electrodes used are not meant for permanent installation, but the measured concentration profiles can give an indication of where, and at what depths, *e.g.* water suction samplers should be installed. Continuous logging of dissolved oxygen is possible using optical type oxygen sensors, yet such devices are still relatively new on the market.

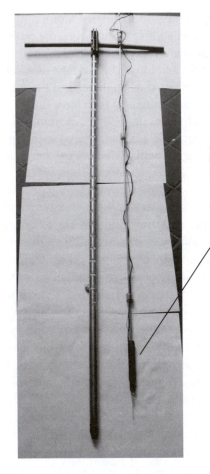

Figure 1 *Equipment used for measuring dissolved oxygen and pH profiles from the soil surface. On the left is shown, the specially-designed steel probe (with white markings for each 5 cm), with the inner rod which has the measuring sensor attached. The middle image shows the dissolved oxygen micro-sensor with protective hypodermic needle. The image to the right shows the pH electrode*

Unfortunately, electrodes for *in situ* measurements are not available for all parameters yet, and the price of the electrodes may be prohibitive. Instead, analysis of water samples from dipwells or piezometers is a reasonable alternative if precautions are taken during sampling. It is essential to purge them prior to sampling, in order to minimise problems with oxygen uptake and oxidation of reduced species in any stagnant water. However, purging alone does not solve all problems related to dipwells. It may actually lead to instantly increased oxygen concentrations in the water. This is probably due to the large surface area of the water trickling into the emptied dipwell, increasing the atmospheric contact and oxygen uptake. Within a few hours or days the oxygen is used by microorganisms or reduced species in the water. A similar pattern has been observed when monitoring the redox potential (Eh) after purging a dipwell. Apart from biasing the oxygen measurements this instant oxygen uptake will, of course, influence the concentrations of other species over time. One of the most sensitive species is dissolved iron: at full oxygen saturation, 20°C and neutral pH (7) it takes only 30 min before 50% of the dissolved Fe^{2+} is oxidised and precipitated as iron(III)hydroxides.

To completely avoid these problems it is necessary to work in an inert atmosphere during purging and handling of samples. As for purging, a continuous flow of nitrogen gas in the headspace of the dipwell can effectively hinder the instant oxygen uptake, and handling of samples in glove boxes in an inert atmosphere is a standard practice when working with marine sediments. This is the ideal solution, but it must also be admitted that such procedures make the sampling and analysis a lot more tedious. If the highest precision is not required, fast sample handling after purging of the dipwells may reduce the error to an acceptable level for all species but oxygen itself.

Following purging, a water sample must be retrieved with a syringe and tube or suitable pumping system. PVC tubing should be used rather than silicon tubing due to its lower oxygen permeability. Water samples should be filtered using 0.45 μm filters, as this by tradition is a standard filter size for distinguishing between dissolved and particulate species in environmental science. All sample handling should be performed as quickly as possible and with minimum contact to atmospheric air – the most sensitive species should be analysed or "fixed" within half an hour of sampling. Ideally samples should be analysed in the field using suitable electrodes or spectrophotometric methods designed for field analysis of water quality parameters. However, if such equipment is not available samples can be fixed and sent away to the laboratory.

2.3 Redox Potential (Eh)

The extent of reducing ion species present can be collectively measured as the Eh. Oxidation involves the loss of electrons, while the process of reduction can be viewed as the gaining of electrons. In order to maintain electrical neutrality overall, it is clear that the oxidation of one species must be accompanied by the reduction of another somewhere in the system: hence, the concept of redox – the simultaneous occurrence of reduction and oxidation. In practice, a chemically inert electrode such as platinum or gold is used, which can transfer electrons to or from the environment. The potential developed is measured relative to a standard cell which itself is calibrated relative to the standard hydrogen electrode. Although measurement of Eh can be problematic, relatively stable measurements can be obtained using permanently installed electrodes of platinum or gold whereas "spot" measurements should be avoided due to unstable and unreliable results. The redox electrodes are connected through a high-impedance voltmeter or pH meter to a reference electrode (such as a RE-5C copper/copper sulfate reference from McMiller Co., which is specifically designed for field use) and readings taken immediately, discarding all results showing strong drift. The necessary impedance of the meter depends on the surface area of the electrode, the required precision of the measurement and the exchange current density of the system studied. The currents involved in the establishment of Eh in the soil are very small, so the potential can easily change (drift) if the current drawn through the measuring circuit is too large. The widespread practice of waiting for a drifting signal to become "stable" before taking the reading is therefore considered a dangerous approach – it cannot be excluded that the drift in some cases is simply due to the measurement itself, changing the potential towards a new equilibrium that has nothing to do with the equilibrium existing in the soil when the redox electrode is not connected! Simple field tests showed that the use of a voltmeter with too low impedance for just 1 min can change the potential of a platinum electrode by several 100 mV, and that it takes several hours before reliable measurements with a high-impedance meter can be made on that electrode again. This short-term drift, caused by the measurement itself, should not be confused with the long-term drift of permanently installed electrodes; the latter is (among other things) probably due to a slow oxygen desorption from the electrode surface and means that it can take several days after installation before the electrode is in equilibrium with its surroundings.

This means that spot measurements, *e.g.* in dipwells or directly in the soil, cannot be compared with results from permanently installed electrodes and that the absolute values from such spot measurements are probably of little use. Experience has shown that redox electrodes can be left in the soil for at least 5 years without major problems.

For most applications platinum has become the preferred material for redox electrodes. However, there is no evidence to suggest that it should actually give better results than gold when measuring Eh in soil with permanently installed electrodes. It is therefore considered that gold electrodes are a good alternative to platinum due to their lower price and lower tendency to form oxides and possibly also sulfides.

2.4 pH

pH is one of the most frequently measured soil parameters, and probably the most indicative single factor for soil functioning and processes. In an archaeological context, pH is a key parameter for the preservation of archaeological remains of several different materials, *e.g.* bone, shells, iron, bronze and even flint. pH measurements are therefore normally included in monitoring programmes at archaeological sites preserved *in situ* and in reburial experiments.

Several different methods have been used for determining pH in these programmes. Most have focused on pore water in the soil, using either dipwells or piezometers in waterlogged environments. Alternately, suction samplers that allow sampling of pore water from soil above the ground water level. In some studies pH has been measured in soil samples, typically in soil sampled during the installation of monitoring equipment. Direct measurement of pH in the soil has been tried at several waterlogged sites using electrodes installed at a selected depth for 7–10 days.

Some of the problems with pH measurements include the oxidation of reduced soil and water samples by atmospheric oxygen, normally causing acid production and thus a decreased pH in the sample. They also include loss of gaseous species such as carbon dioxide, ammonia or hydrogen sulfide during and after sampling. This is due to the different partial pressures below and above ground – *e.g.* the partial pressure of carbon dioxide is often 10–100 times higher in the soil than in the atmosphere, due to root respiration and decomposition of organic matter. The magnitude of these problems will vary between different methods and sites. For instance, degassing of carbon dioxide during the use of suction samplers can easily lead to pH increases of $\frac{1}{2}$–1 pH units and degassing during drying of soil samples may give deviations as high as 2 pH units. Deviations are also possible when water is standing in a dipwell open to the atmosphere. Theoretically, the ideal solution is to measure the pH *in situ* in the soil rather than taking samples. This approach has recently been evaluated making *in situ* measurements of pH from the soil surface using a specially developed steel probe incorporating an Ion Sensitive Field Effect Transistor (ISFET) electrode, which is a very sturdy, solid-state pH electrode (Figure 1). This method gave reliable results and avoided problems with oxidation and degassing of soil or water samples, and allowed the measurement of quite detailed pH profiles from the soil surface. It can also be recommended to use the ISFET electrode during archaeological excavations as it allows an immediate evaluation of the pH values and burial conditions in open soil profiles with a minimal disturbance of the soil. However, *in situ* measurements with an ISFET electrode should be considered a supplement to, rather than substitution for, the existing methods. For instance, the ISFET electrode used is not suitable for permanent installation.

3 DETERIORATION OF MODERN MATERIALS ANALOGOUS TO ARTEFACTS

Further to monitoring the environment itself, modern materials placed on a site can serve as analogues to the types of archaeological material found. In this way their deterioration can be regularly monitored in order to better understand the processes. Furthermore, these materials serve as a "proxy" indicator of the environment, and can potentially serve as a warning should the environment not be conducive to preservation. Ideally these materials are placed on the site with the minimum amount of disturbance to both site and environment. Sufficient samples should be placed for long-term monitoring; ideally 50–100 years and with samples sufficient for at least triplicate analysis. To this end systems have been developed for placing and retrieving wooden and metal specimens without the need for excavation.

A system designed for placing wooden samples on a site (Figure 2), consists of a carbon fibre pole with a sharpened spike, machined from a solid piece

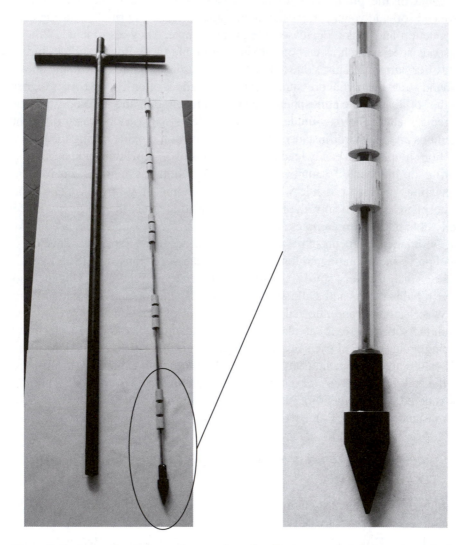

Figure 2 *Equipment for placing modern wood samples. The image to the left shows the steel tube and the carbon fibre pole with mounted specimens. The image to the right shows an expanded view of the tip of the pole and the arrangement of specimens and PVC spacers*

of PVC, fixed at one end. A series of small cylindrical wood samples are threaded onto the pole by means of holes drilled through their longitudinal axis. Spacers of PVC tubing are placed between samples so that their position on the pole corresponds with the depth of the artefacts, or the layers, which are to be studied. Presently pine and balsa samples have been deployed on sites in Denmark. These woods have been chosen because of their relative susceptibility to waterlogging, tendency to degrade and, certainly with pine, the relative ease of understanding the wood ultrastructure and decay patterns produced by microorganisms. Investigations using other wood types are planned. However, it is not necessarily the rate of decay (which is influenced by wood species) that is important in this instance but the types of decay that are seen.

The pole and samples are placed within a metal tube, the diameter of which is slightly larger than the wooden samples yet smaller than the diameter of the base of the spike. The metal tube and pole are then pressed or hammered into the soil to the required depth. At least 30 cm of the pole are left above ground to aid in locating and retrieval. The metal tube is thereafter pulled up leaving the wood samples exposed to the soil conditions. The benefit of using many small samples instead of one continuous wooden pole is that the surface area of the longitudinal axis is greatly increased. Retrieval is achieved by pulling up the carbon fibre pole using a tightly fitting clamp. In waterlogged soils, placing and retrieving wood in this way is relatively straightforward, but the system has yet to be tested on underwater sediments or in drained soils. The extent of deterioration can be determined through changes in density and the types of decay present assessed by various microscopic techniques.

For metals, individual "coupons" are tied to a thin polyamide thread and the end of the coupon placed into a groove cut into the end of a metal rod. The metal rod and coupon are then placed into a hollow steel tube, which has a sharpened spike at one end with a thin slit cut into it to allow the coupons to pass through. The whole assembly is pressed down through the soil to the desired depth, and the thin rod is then pushed down forcing the coupon through the slit. The coupon is released by a slight twist of the metal rod. The steel tube is removed and the metal coupon is left in the soil still connected to the surface by the thread, which is then used to retrieve it after the desired length of time. These coupons are used for the study of weight loss and corrosion products. Other probes have been designed to allow the measurement of corrosion potentials, electrical resistivity (ER) and electrochemical impedance spectroscopy (EIS).

4   WHAT DOES ENVIRONMENTAL MONITORING ALONE TELL US?

Environmental monitoring alone does not tell anything about the actual state of preservation of artefacts preserved *in situ*, and should only represent one part of a "three-pronged" approach to the successful *in situ* preservation of an archaeological site, namely:

(i) environmental monitoring;
(ii) analysis of the deterioration processes of modern materials analogous to archaeological artefacts; and
(iii) analysis of artefacts to understand the processes of their deterioration and their future stability.

This overall approach will be illustrated with results from the site of Nydam.

4.1   Case Study: *In situ* Preservation of the Site of Nydam

Nydam mose is an 11-hectare water meadow located in Southern Jutland, Denmark (UTM coordinates are 546620/6090831 ED50 Zone 32). During the Danish Iron Age it was a shallow freshwater lake into which sacrificial offerings of military equipment were deposited on at least five different occasions in the period from approximately 200 to 500 AD. Excavations have taken place on several occasions since 1859, and the latest campaign, from 1989 to 1997, yielded more than 15,000 archaeological artefacts within an area of only $600 \, m^2$ (Figure 3). Owing to the volume of finds it was decided in 1997 to stop further excavation campaigns and preserve the site *in situ*.

Figure 3 *The range and richness of finds in Nydam. Spears, arrows, swords, shields and bosses*

The archaeological remains are found primarily in two layers, approximately 0.80–1.25 m below the surface (Figure 4). The uppermost finds are lying in peat, whereas the deeper lying finds are in a carbonate-rich gyttja.

Wood and iron are the most abundant materials, but also artefacts of bone, leather, amber, glass, bronze, and precious metals are found. Following on from the period of extensive environmental monitoring at the site between 1996 and the present, it could be seen that the water level has been well above the find layers and the seasonal fluctuation in water level is modest at approximately ±40 cm. The meadow is slightly acidic (pH 6–7) and below the root zone there is very little vertical variation. A good buffer capacity in the peat and gyttja layers helps to minimise fluctuations. Dissolved oxygen content is <0.1 mg L^{-1} and the Eh (measured with permanently installed gold electrodes) is stable at −200 mV, indicating reducing conditions which is supported by the levels of reduced chemical species measured in pore water from the site. Ideal conditions for preservation, but what does all this information say about the state of preservation and importantly the future state of preservation of artefacts preserved *in situ*? Research on artefacts themselves has focused primarily on the deterioration of wooden and iron artefacts. These form the most abundant groups of material encountered, and, based on the

Figure 4 *The find layers typically found in Nydam. Finds are primarily found in two layers – the Upper find layer (to the left of A) and the Lower find layer (to the left of B). The finds in the upper layers are surrounded by peat whereas those in the lower layers tend to be surrounded by gyttja*

condition of finds from excavations in the 1990s, are under most threat of deterioration if left *in situ*.

4.2   Deterioration and Protection of Wooden Artefacts

From the amount of wooden finds still in Nydam, the environmental conditions can be considered conducive to the preservation of wood. However, archaeologists working at Nydam were concerned that the finds excavated during the 1990s were in a worse state of preservation than those excavated in the 19th century. This concern is based on three observations. First, the last 10–15 years has seen the growth of "horsetails" (*Equisetum*) over Nydam. Although they seem quite small and innocuous on the surface, their root system can grow down one to two metres. This not only brings oxygen down to the find layer but, more drastically, the roots actually penetrate the artefacts.

Second, when the artefacts are viewed by eye, there is a distinct difference between many of the artefacts in the different find layers. Many in the upper find layers have a roughened, gnarled surface appearance whereas those in the lower find layers often retain many surface details (Figure 5). Third, density measurements taken on the planking from the boats excavated in 1864 showed that the average density was $460 \, \text{kg m}^{-3}$, whereas smaller artefacts excavated more recently show an average density of $100 \, \text{kg m}^{-3}$. Monitoring alone could not substantiate these observations. Therefore, a series of artefacts, representative of those in the different find layers, were analysed to assess their state of preservation through determination of their density and cell wall components (cellulose and lignin). DNA was also extracted to try to identify the microorganisms, which had caused deterioration. Macroscopic and microscopic analysis of artefacts enabled background information on the post-depositional processes of deterioration and the present state of preservation to be investigated. Furthermore, experiments looking at the deterioration of wood by basidiomycete fungi and the tendency for artefacts to collapse, showed what could happen should the water level in Nydam fall below that of the find layers.

Macroscopic analysis of artefacts from the upper find layers in Nydam showed that those artefacts, which, from their surface details, appeared to be in a poorer state of preservation, were actually degraded prior to their incorporation in the meadow. Through the analysis of the surface of the artefacts it was possible to elucidate that the most likely cause of the deterioration was through the activity of the wood-boring beetle *Nacerda melanura*, which is often found to cause deterioration of wood that lies exposed in, and around, wet environments. This would suggest that the artefacts displaying such surface details had lain exposed, perhaps floating on the surface of the lake, prior to being incorporated in the bog. This observation was also supported by microscopic analysis of their structure. They also showed a dark ring

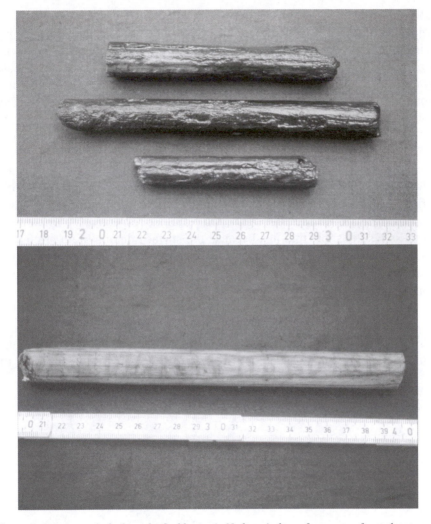

Figure 5 *Ash spear shafts from the find layers in Nydam. A shows fragments of an ash-spear shaft from the upper find layer with a typically gnarled surface. B is of an ash-spear shaft from the lower find layer with runic style markings well preserved on the surface*

around the outer 1–2 mm, which was seen to be an effect of degradation by soft rot fungi, which are known to require oxygen for their respiration. Finds in the lower find layer did not show signs of soft rot and only signs of deterioration characteristic of erosion bacteria, indicating that these artefacts had been quickly incorporated into an anoxic environment. However, in combination with the physical and chemical analysis and environmental monitoring results, it is unlikely that the artefacts will degrade further as long as the environment remains stable. Microscopic analysis with polarising light showed that in all artefacts, regardless of their spatial position in the site, the secondary cell wall and most of the primary cell wall had been utilised by micro-organisms. All that remained was the compound middle lamella, its shape being retained because the degraded cell lumen was filled with water and bacterial degradation products (Figure 6). These observations were reflected by the bulk density and cell wall component analyses. The bulk density of all small artefacts analysed was between 90 and 110 kg m^{-3} regardless of the spatial position within Nydam. There was no "density gradient" measured within the artefacts as is often seen in archaeological artefacts. This would seem to be the lowest density of wood, which can be achieved – regardless of age, environment and wood species, and reflects the density of the compound middle lamella. This was supported by the cellulose and lignin results, which reflect the expected amounts to be found in the compound middle lamella (approximately 60% and 20% lignin and cellulose respectively). Importantly, all these results show that just because the surface details of an artefact are poorly preserved from an archaeological perspective, does not mean that the artefact is in a worse state of preservation when considered from a conserva-tion perspective.

It is highly unlikely that there will be further microbial degradation of the smaller artefacts if the environment remains waterlogged. Yet, with the advent and use of molecular biological techniques to identify wood-degrading microorganisms this may be confirmed or refuted. The implications of this

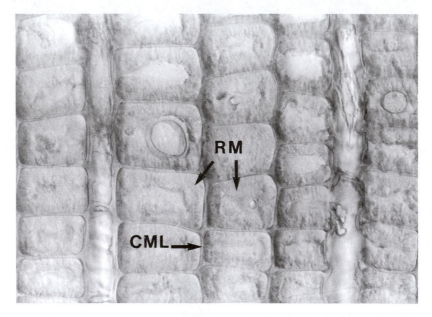

Figure 6 *Deterioration of a pine arrow shaft by Erosion Bacteria. Secondary cell wall (RM) converted into residual material. Compound Middle Lamellae (CML) physically retains the dimensions of the cell*

are especially important if larger oak timbers, which from existing data are potentially not as degraded as the smaller artefacts, are present on the site. However, if the water level in Nydam should fall below that of the find layers, there may be the risk of fungal deterioration. Fungal decay can be divided into three groups: white, brown and soft rots. White and brown rots (Basidiomycetes) generally require high levels of oxygen and non-waterlogged conditions in order to decay wood. They predominate in well-drained and aerated soils and lead to the relatively rapid and total destruction of wood. Soft rot fungi (ascomycetes and fungi imperfecti) appear to be tolerant of lower oxygen concentrations and waterlogged conditions in soils. Experiments were conducted to simulate the effect of drainage by inoculating pieces of a sterilized ash spear shaft from Nydam with strains of white and brown rot fungi. It was found that after 8 weeks, white rot was able to utilise the already degraded wood tissues, whereas brown rot showed no effect.

Although these results have implications for the long-term preservation and/or deterioration of archaeological wooden artefacts, the results of the collapse experiments show that the majority of important archaeological information (surface details) would be lost as soon as the artefacts begin to dry (<99% relative humidity). Thus, it is important to know whether the artefacts are so degraded that they are likely to suffer collapse when considering their *in situ* preservation. Similarly, it is important to know whether an artefact has suffered collapse prior to determining bulk density as, if so, artificially high results will be obtained, giving the impression that artefacts are better preserved than they really are. It appears that as long as the water level in Nydam remains stable there will be limited deterioration from micro- and macro-organisms. Modern wood samples were placed on the site in 2002 and results after 1-year of exposure show no signs of deterioration. Yet, this may simply be due to the short exposure time. The major threat still present on the site appears to be the roots of the *Equisetum*. Pilot studies to prevent their growth are ongoing with attempts to physically cut the horsetails periodically (labour intensive and thus costly). A further method is placing geo-textile over the parts of the site and monitoring to see if it prevents their growth. The environment under the geo-textile is also being monitored to see if it is changing as a result of this.

4.3 Deterioration of Iron Artefacts

Analysis of metal artefacts tells us about their state of preservation, and analysis of any covering corrosion products informs us about the post-depositional corrosion processes and their thermodynamic stability. This in turn enables us to forecast the changes in the environment that may be detrimental to their future preservation.

The state of preservation of iron artefacts from Nydam showed a very varying picture, with some objects being completely converted into corrosion products, while the original surface of others was still visible. During the excavation, it was noted on some occasions that well-preserved and heavily deteriorated objects were lying just a few centimetres from each other. In order to evaluate this further, a group of artefacts (spear and lance heads) were selected for a closer study of their state of preservation. The spear and lance heads were X-rayed, and the corrosion depth was measured at 20–30 positions on each artefact and averaged. The measured corrosion depths were correlated to the exact find position using a Geographical Information System (Figure 7).

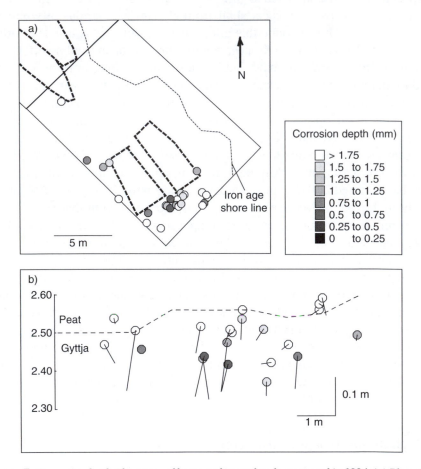

Figure 7 *Corrosion depth of a group of lance and spear heads excavated in 1994. (a) Plan view, where solid lines indicate modern excavations and dashed lines show excavations during the 19th century. (b) Vertical profile (seen from SE) where dashed line indicates interface between peat and gyttja. Solid line at each point represent a projection of the lance or spear head to the vertical view plane, so long steep lines indicate artefacts deposited in a steep angle. Numbers to the left are metres above sea level*

The results indicate that the overall deterioration pattern observed today is a function of the first few years after deposition. It seems that artefacts that have been thrust down into the anoxic peat layers are less deteriorated than artefacts that have been sacrificed on the mire surface. The latter have lain exposed in an oxygen-rich environment for a period of time before being finally overgrown and embedded in an anoxic environment, and these differences from 1500 to 1800 years ago are still observable today.

The corrosion scales on the archaeological artefacts have been shown to consist primarily of siderite ($FeCO_3$) visible as large yellow–brownish crystals on the surface of some of the artefacts. Analysis of cross-sections of a few artefacts also showed that the inner corrosion layers consisted of siderite with some magnetite or maghemite. Importantly, only very low contents of sulfur were found, indicating that corrosion by sulfate-reducing bacteria is not a problem at this site. Siderite was also found on modern iron samples that had been exposed for a few years in the soil at Nydam.

Siderite is a common mineral in mires, where it is formed through microbiological reduction of iron oxides in the environment. This mechanism may explain its occurrence on artefacts that have lain exposed on the mire surface for a period of time. In these conditions they will be quickly covered by a layer of iron oxides, which will subsequently be reduced to siderite after being overgrown and embedded in an anoxic environment. However, for other artefacts (and modern samples) that have been placed directly under anoxic conditions the siderite must have formed directly from the metallic iron, and here it is still unclear exactly what cathodic reaction is responsible for the oxidation of iron. A Pourbaix diagram based on the actual soil conditions at Nydam is shown in Figure 8. The hatched area in the Pourbaix diagram demonstrates that the pH values found at Nydam are close to the lower limit for siderite stability, so the soil pH is monitored intensively to be sure that no acidification takes place.

Weight loss has been measured for more than 40 modern iron coupons; each exposed for approximately 2 years in the waterlogged peat and gyttja layers in Nydam. The results have demonstrated a close correlation between archaeological excavations in the area and measured corrosion rates (Figure 9).

The highest rates of corrosion were found near the archaeological excavations, where the water level had been lowered by pumping during the excavations. In these areas temporary changes in the water chemistry were also measured, such as an increased concentration of nutrients and sulfate, and a slight lowering of the pH. It must be emphasised that on both occasions the coupons were placed in the soil several months after excavations were finished, and all coupons have constantly been below the level of the water table. It remains to be shown whether the archaeological artefacts in the area have also been affected by the excavations.

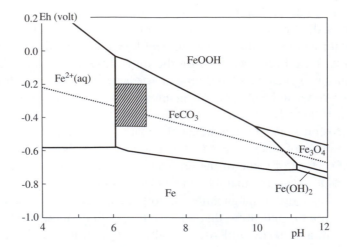

Figure 8 *Pourbaix diagram for iron in an environment with 10^{-2} M carbonate, and 2×10^{-4} M dissolved iron at a temperature of 10°C, generated by the software HSC chemistry 5.0 from Outokumpu. The thermodynamic data for the different corrosion products are compiled by the French Agency for Nuclear Waste Management (ANDRA). Hatched area shows the pH values and corrosion potentials measured in situ at Nydam. Dotted line shows the stability limit for water*

By combining the information from environmental monitoring, modern coupons and archaeological artefacts it is possible to give an overall evaluation of the preservation conditions for iron in Nydam. The majority of deterioration of the artefacts probably took place in the first few years after sacrifice (Figure 7), and we consider the current conditions as conducive to the *in situ* preservation of the artefacts. However, the water level and soil pH have to be monitored also in the future. It seems that even a slight lowering of the soil pH may cause a dissolution of the corrosion products covering the artefacts (Figure 8), and from the modern coupons it seems that even a temporary lowering of the water table may have an adverse effect on the preservation (Figure 9).

5 CONCLUSIONS

In situ preservation is a form of preventive conservation, but absolute preservation *in situ* is not achievable. All sites are dynamic and deterioration of archaeological materials will continue, *albeit* at slow, and often imperceptible rates. However, environmental conditions favourable to the preservation of archaeological sites, and artefacts, can be assured by a "three pronged" sequence of environmental monitoring, study of the deterioration of analogous modern materials placed on a site, and the state of preservation of artefacts themselves.

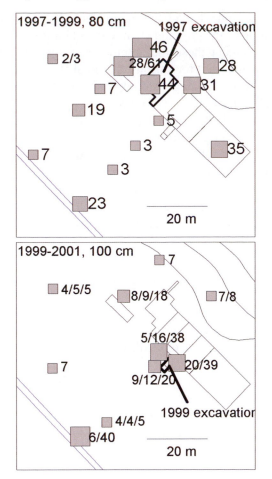

Figure 9 *Corrosion rates on modern iron coupons. One set of coupons were buried in October 1997 at 80 cm depth and retrieved in May 1999. The other set was buried in November 1999 at 100 cm depth and retrieved in June 2001. All numbers are in* $\mu m \times y^{-1}$

■: *0–10,* ■: *10–20,* ■: *20–30,* ■: *30–40,* ■: *>40* $\mu m \times y^{-1}$

One of the fundamental parameters to the preservation of archaeological sites in waterlogged conditions is the water level – this factor controls many of the processes ongoing in the soil. Changes in water level can be monitored using dipwells/peizometers with simple measuring techniques or with data-loggers. Along with the water level, the dissolved oxygen content of the water/soil affects the processes of deterioration of organic materials, such as wood, and the corrosion of metals. Ideally this parameter, along with other parameters such as the pH and the Eh, should be measured *in situ* with special probes, in order to get a better understanding of the oxidising/reducing nature of the environment, and the acidity/basicity of a site. From these results an indication of the ongoing chemical and biological processes can be obtained. To supplement these measurements water samples can be taken from dipwells and analysed for various chemical species.

Further to monitoring, analogous modern materials can be placed on a site. The study of their deterioration will serve as a "proxy" indicator of the environment, giving an indication of what is presently happening on the site and, importantly, whether mitigation strategies are working.

Finally, analysis of artefacts themselves should be carried out in order to assess their state of preservation. By understanding this and determining the likely pathways of deterioration the stability of a site can be assessed. Passive or active mitigation strategies can then be implemented (with further monitoring) to ensure that sites, or artefacts, preserved *in situ* are protected for the future.

REFERENCES AND FURTHER READING

Further references to publications on research into *in situ* preservation of archaeological artefacts and sites are available from the authors or from the following home page: www.natmus.dk/sw8878.asp

R. Brunning, D. Hogan, J. Jones, M. Jones, E. Maltby, M. Robinson and V. Straker, Saving the sweet track, The in situ preservation of a neolithic wooden trackway, *Conserv. Manage. Archaeol. Sites*, 2000, **4**, 3–20.

J. Buffle and G. C. Horvai, (eds), *In situ Monitoring of Aquatic Systems: Chemical Analysis and Speciation*, Chichester, Wiley, 2001.

C. Caple, Defining a reburial environment: Research problems characterising waterlogged anoxic environments, *Proceedings of the 5th ICOM group on wet organic archaeological materials conference, Portland, Maine*, in P. Hoffmann (ed.), Bremerhafen: ICOM, 1994, 407–421.

C. Caple and D. Dungworth, Waterlogged anoxic archaeological burial environments, *English Heritage Ancient Monuments Laboratory Report*, 1998, 22/98, 1–137.

A. Christine Helms, A. Camillo Martiny, J. Hofman-Bang, B. Ahring and M. Kilstrup, Identification of bacterial cultures from archaeological wood using molecular biological techniques, *Int. Biodeter. biodegr.*, 2004, **53**(2), 79–88. (This Journal also contains good information on the biodeterioration and corrosion of cultural materials.)

M. Corfield, Preventive conservation for archaeological sites, *Archaeological Conservation and its Consequences. Preprints of the Contributions to the Copenhagen Congress*, in A. Roy and P. Smith, (eds), *26–30 August 1996*, London: IIC, 1996, 3237.

M. Corfield, P. Hinton, T. Nixon and M. Pollard. (eds), Preservation of archaeological remains in situ, *Proceedings of Conference held at the Museum of London, April 1996*, Museum of London Archaeology Service, London, 1998.

D.V. Hogan, P. Simpson, A.M. Jones and D. Maltby, Development of a protocol for the reburial of organic archaeological remains. *Proceedings of the 8th ICOM Group on Wet Organic Archaeological Materials Conference, Stockholm, 11–15 June 2001*, in P. Hoffmann, J.A. Spriggs, T. Grant, C. Cook and A. Recht, (eds), Bremerhaven: ICOM 6-11-2001, 2002, 187–212. (Many articles on the deterioration of organic materials and *in situ* preservation of archaeological sites and artefacts can be found in other volumes of the WOAM Group.)

G. Kirk, *The Biogeochemistry of Submerged Soils*, Chichester, Wiley, 2004.

T. Nixon, (ed), *Preserving Archaeological Remains In situ. Proceedings of the 2nd Conference 12–14 September 2001*, Museum of London Archaeology Service, London, 2004.

R.M. Rowell and R.J. Barbour (eds), *Archaeological Wood Properties, Chemistry and Preservation, Advances in Chemistry Series 225*, American Chemical Society, Washington, 1990, 158–161.

W. Scharff, C. Arnold, W. Gerwin, I. Huesmann, K. Menzel, A. Pötzsch, E. Tolksdorf-Lienemann and A. Tröller-Reimer *Schutz archäologischer Funde aus Metall vor immissionsbedingter Schädigung*, Stuttgart: Konrad Theiss Verlag, 2000.

J. Schüring, H.D. Schulz, W.R. Fischer, J. Böttcher and W.H.M. Duijnisveld, *Redox: Fundamentals, Processes and Applications*, Berlin, Springer, 2000.

B. Sørensen and D. Gregory, *In situ* preservation of artifacts in Nydam Mose, in *Proceedings of the International Conference of Metals Conservation*, W. Mourey and L. Robbiola (eds), Draguignan-Figanières 1998, 94–99. (Many articles on the corrosion of metals and their *in situ* preservation can be found in other volumes of the ICOM-CC Metals Group.)

W. Stumm and J.J. Morgan, Aquatic Chemistry: An Introduction Emphasizing Chemical Equilibria in Natural waters, New York, Wiley, 1981.

Subject Index